OIL IN THE EARTH

Oil
In the Earth

by

WALLACE E. PRATT
Vice-President, Standard Oil Company (New Jersey)

AN ELEMENTARY DISCUSSION OF A PRIME NATURAL RESOURCE
BASED ON FOUR LECTURES DELIVERED BEFORE THE STUDENTS OF
THE DEPARTMENT OF GEOLOGY, UNIVERSITY OF KANSAS, MARCH
17 TO 19, 1941. FOR THE SECOND EDITION THE ORIGINAL SUB-
JECT MATTER HAS BEEN SLIGHTLY REVISED, REARRANGED AND
AMENDED.

UNIVERSITY OF KANSAS PRESS
LAWRENCE
1943

PRINTED BY THE UNIVERSITY OF KANSAS PRESS
LAWRENCE, KANSAS

TO
IRIS CALDERHEAD

"It is obvious that the total amount of petroleum in the rocks underlying the surface . . . . is large beyond computation."

(ORTON, 1888.)

CONTENTS

I.   THIS IS OIL 1

II.   WHERE OIL IS 27

III.   WHO FINDS OIL AND HOW . 53

IV.   WHOSE OIL IS IT 77

I

THIS IS OIL

Some of the subject matter of these lectures has already appeared in print in different papers published in the Bulletin of the American Association of Petroleum Geologists, Tulsa, Oklahoma; the Oil Weekly, Houston, Texas; and Mining and Metallurgy, American Institute of Mining Engineers, New York. This material is reproduced here with the permission of these publishers, whose courtesy is gratefully acknowledged.

I

Introduction

WHY should former students be invited to address the Department of Geology? I confess my pleasure at the privilege you have accorded me; but the lecturer's pleasure hardly justifies the custom of bringing back an old grad and inflicting him upon the student body.

It might be supposed that lectures by professional scientists would tend to diminish the existing lag between the acquisition of new knowledge in the field and its incorporation into the curriculum of our technical schools. Uniformly in our social order a long period intervenes between the discovery of new facts at the frontiers of industry and the effective presentation of these facts at the established educational centers. But the occasional lecturer cannot contribute much toward eliminating this handicap to progress. The remedy lies elsewhere: probably in fundamental reform of the relationship between science and industry.

I cannot hope to present any facts or theories on the nature of oil in the earth, or of the geology of oil fields which are not already well known to you from your regular lecturers, your references and text books. At best, I can only repeat a part of what your instructors have already said with greater clarity and accuracy. The single contribution I can make to your professional equipment, I believe, is to tell you something of my own experience as a geologist in the field of oil-finding.

In drawing heavily on my personal experience I am adopting a strategy long known to occasional lecturers. Henry D. Thoreau,[1] who in another age might well have been a geologist, defends the logic of such a practice: "I should not talk so much about myself," he says, "if there were anybody else whom I knew as well . . . . I take it for granted when I am invited to lecture anywhere — for I have had a little experience in that business — that there is a desire to hear what I think on some subject . . . . **and I resolve, accordingly, that I will give them a strong dose of myself; they have sent for me . . . . they shall have me though I bore them beyond all precedent.**"

Thoreau's method releases a flood of pronouns, first person, singular, but I adopt it, nevertheless, to recount my own experience among oil fields. As a boy, nearly forty years ago, I came to K. U. bent upon a very serious inquiry; I thought to learn how the eyes of wisdom viewed the world and the activities open to young men. For me it was to be a

[1] Life without Principle. The Writings of Henry Thoreau, page 711, Modern Library, New York.

tremendous excursion into the domain of reflective thought. Almost at once it seemed my course began somehow to shape itself and to carry me along with it; from liberal arts to engineering, to the sciences and then into the special field of geology. More than thirty years ago I left K. U. to go to work in applied geology. Through great good fortune, as I count it, I have been enabled to devote my life to the profession of economic geology, largely in the search for oil in the earth. Let me speak to you then of my observations on the modern quest of mankind for new oil fields.

When I left K. U. in 1909, Kansas was producing only 3,000 barrels of oil per day; less than 3 per cent of the production of her neighbor, Oklahoma, and hardly more than 1 per cent of Kansas' present production. The State Geologist, Erasmus Haworth (we affectionately called him "Daddy"), assisted by a group of students, was busy getting out Volume IX of the Kansas University Geological Survey on the subject of the "Oil and Gas Resources of Kansas"; yet, in fact, those resources bulked rather small in the eyes of the oil industry. In retrospect, some of the ideas we then held of the occurrences of oil appear now to be of doubtful validity. We set it down in Volume IX that "where valuable production has been obtained there have been practically no seeps of either oil or gas. If oil and gas are escaping at the surface", we wrote, "it is a sure indication they do not exist in the depths below . . . . in very large quantities." Experience has led us to discard this thesis from our store of accepted principles. Another "gospel truth" of that time now thoroughly discredited was that the "Mississippi Lime" was the absolute floor of nature's storehouse for oil under the plains of Kansas. The unconformity at the top of the Mississippian series we then thought marked the extreme depth at which we could hope to find oil. To have suggested the possibility of encountering oil in a lower horizon, the now copiously productive Ordovician, for example, would have stamped one as a crackpot.

Experience has invalidated other firm beliefs of those days. We thought that the sedimentary section in Kansas was profoundly thick. We would probably have guessed that the vertical distance to the crystalline floor under central Kansas was nearly equal to the horizontal distance to the crystalline core of the Rocky Mountains. When in 1915 the driller of a "wild-cat" well reported granite at a depth of 1100 feet, or so, right in the middle of the state, we were not only skeptical; we were indignant. We denied that the well had granite; and when the driller, under our own supervision, bailed out of the well fragments of beautiful pink granite, we charged that he had planted the granite in there himself! So little were we prepared for anything like the now classic buried Nehama Mountains, traversing the width of our state just under its surface: still less could we have conceived that such a buried mountain chain, under the sediments spread over it by encroaching seas, might be fringed with porous reservoirs filled with oil.

A little later I discussed with the chief geologist of a large oil-producing company the prospect for new oil fields in Kansas. He was skeptical. He pointed out that the rocks beneath Kansas consist largely of limestone and shale whereas oil is usually found in a stratigraphic sequence of sandstone and shale. Accordingly he would rate the prospect in Kansas as uninviting. Only a few months later the famous Eldorado Field east of Wichita was opened up, and reviewing our discussion in the light of this discovery I concluded there are important exceptions to perfectly valid general principles in geology.

Such episodes illustrate a reciprocal stimulation between science and industry. This important effect is broader, to be sure, than the relationship between geology and oil-finding, but it has operated to the extraordinary benefit of these particular enterprises. The Kansas wild-catter, drilling for oil and encountering a buried mountain range, not only astounded the geologist; he uncovered a foundation for the revision and expansion of the geologist's whole conception of the occurrence of oil fields. This new conception, promptly reduced to practice by the geologist, soon enabled him to identify hitherto unsuspected oil fields for his client, the wild-catter.

As geologists we like to emphasize the contribution we make to the oil industry. Less emphasis on the contribution of oil producers to our knowledge of geology has found its way into print. If we strike a balance in terms of broad social values we may find that science, not industry, is the debtor, notwithstanding the magnificent achievement of geologists in finding oil. Those of us who can recall the commonly accepted conception of the geology of Kansas before we knew what wells drilled for oil have since revealed to us, might agree that our effort to make geology serve industry has really benefitted geology more than industry. The bread we have cast upon the waters has come back to us many-fold. We have learned far more than we originally had to impart. Moreover, our new knowledge is worth more to society at large than the new oil we helped the wild-catter to find.

The discovery and production of oil is a geological enterprise. It is of the nature of oil in the earth that, so intimately are the two associated, we develop our oil resources most efficiently if we know perfectly the outer shell (at least) of the earth. Our effort to discover new oil fields has been handicapped by our lack of knowledge of the earth in which oil fields are concealed. So little have we known that our actions have sometimes been ludicrous. I abandoned Kansas in my quest of oil fields: but while I searched in far places, Kansas' own production increased nearly a hundred-fold. It is a little humiliating to the geologist, engaged over thirty years in search of oil, half way around the earth, to learn upon his return at length to his birthplace that oil is gushing from all the farms in the neighborhood.

Just what is this oil which gushes from the earth? The subject of these lectures might have been expressed by the single word "petroleum". Petroleum, literally rock oil, is the generic term for all natural hydrocarbons, gaseous, liquid and solid (except coals). I have used the longer title because I hope to convey an even broader conception. Petroleum is more than rock oil; it is "earth oil"; it is independent, in a sense, of the rocks in which we find it; but it is an inevitable result of fundamental earth processes; processes so typical that they have been repeated in each successive cycle of earth history. The Greeks saw petroleum as an attribute of the rocks but we have learned that the association with a particular rock is only an incident; petroleum derives from the earth at large. I believe that oil in the earth is far more abundant and far more widely distributed than is generally realized. Oil is a normal constituent of unmetamorphosed marine rocks of near-shore origin. Rocks of this character comprise nearly forty per cent of the total land surface of the earth. The veneer of marine sediments that comes with time to be incorporated into the earth's crust is the native habitat of oil. Oil is a creature of the direct action of common earth forces on common earth materials.

Oil is a complex mixture of hydrocarbons. Gas is also a complex mixture of hydrocarbons. The solids, asphalt and bitumen, are also complex mixtures of hydrocarbons. Hydrocarbons consist of complex molecules made up of varying combinations of the atoms of two elements, hydrogen and carbon, which are present on (and in) the earth in great abundance. There is a multitude of hydrocarbons varying in a uniform series from the compounds of large molecules which comprise the solid, asphalt, through the intermediate-sized molecules of oil to the smallest molecules, gaseous at ordinary temperatures and pressures. Oil, gas and the solid hydrocarbons are closely related, then, and changes in temperature and pressure readily transform one into another. Both gas and solid hydrocarbons are soluble in oil. Oil in its natural state in the earth often contains many times its own volume of gas in solution. Oil which finds its way to the surface of the earth and is exposed to the atmosphere usually loses its lighter, smaller molecules through evaporation and thereupon assumes the solid form.

THIS IS OIL

NATURE OF OIL POOLS

In a long series of geologic events and only as a single stage in the series, commercially important accumulations of oil in the earth, our so-called "oil pools" or "oil fields", develop. Oil pools are found almost invariably in marine rocks, those successive layers of sand and mud deposited as sediments over the floors of former seas. With the accumulation of the overlying load of sediment the deeply-buried layers gradually harden into rock. As compaction proceeds under this pressure the muds become dense and impervious to fluids. But in the sandstones remain pore spaces and these continue to be occupied by the fluids, principally sea water, in which the component grains were laid down. Among these fluids are minute bubbles of oil and gas, products of the alteration of organic matter buried with the mud at the time of its deposition on the sea floor. These bubbles of oil and gas, together with much larger volumes of sea water, move under the pressure of compaction from the muds to the adjacent sand layers. The oil and gas bubble upward through the sea water in the pores of the sand layer until their upward movement is stopped by the next overlying layer of impervious mud. Now the oil and gas can move only laterally along the roof of the sand layer on top of the brine which permeates all the surrounding pore space. As lateral movement through the sand proceeds the bubbles coalesce to become larger and larger. Eventually they encounter some dome-shaped concavity or other irregularity in the impervious roof; in these "traps" they collect and there they are confined by the hydraulic pressure of the underlying brine on which they float. Such trapped accumulations of oil and gas we designate as oil pools.

An oil pool is a closed system or equilibrium of forces. Gas is readily compressible; in the underground reservoirs it is compressed by the hydrostatic head of the salt water which seals the oil pool in its confining trap. Geologists and engineers have learned to recognize this equilibrium of forces and gradually the industry at large learns a new economy; to substitute the work of these natural forces within the earth for the labor of man on top of the earth to recover oil and lift it to the surface. There are further implications in these closed systems of forces and, once comprehended, these implications become significant as guide posts to the discovery of new oil fields.

Not only the ordinary physical forces, pressure, friction, capillarity and gravity are involved in the balanced system of an oil pool in the earth's crust; there are also forces which change the relations of the elements one to another and control the union or recombination of the elements into new compounds. We have learned to utilize some of these forces in our stills on top of the ground; cracking, polymerization and hydrogenation, for example. Present also in natural reservoirs in the earth's crust, as well as in our equipment above-ground, are the catalysts that promote chemical reaction so effectively. In oil pools one hydrocarbon is constantly changing to another. Large molecules are cracked; smaller molecules are polymerized; hydrogen is added and taken away. If in the earth these processes go forward slowly they go forward none the less surely. When physical conditions change, not only do pressure and temperature respond, but the equilibrium between the various hydrocarbon molecules that comprise oil and gas is destroyed and a new equilibrium tends to establish itself on a basis that reflects the new temperature-pressure relation. These changes which take place in the oil pool are the changes which we also induce in the stills of our refineries above ground. Thus does man "hold a mirror up to nature."

The earth's crust is a dynamic system, prone to cyclic change through geologic time. The profound deformation which these changes inflict on the rock layers often destroys a porous reservoir and dissipates the oil and gas contained within it. So common is this phenomenon that it may be said that oil is only transitory in the earth's crust. Oil pools are evanescent; they seldom survive profound geologic change. Only in stable sectors of the earth's crust do oil pools remain intact through one geologic cycle after another. Accordingly, most of our oil in the earth is in the geologically young rock layers; comparatively little remains in the older rocks.

The inherently transitory nature of oil fields becomes more comprehensible if we realize one attribute of the fundamental character of oil in the earth. We can more readily understand that oil fields cannot survive drastic deformation of the earth's crust when we know that the major component of an oil field is gas, with all its propensity for movement, escape and dissipation in response to pressure. Oil fields are essentially gas fields. While we record the volume of our discoveries in terms of barrels of oil, what we really find is gas with which is associated a subordinate quantity of oil. Our search for oil in the United States over the last ten years has brought us greater reserves of energy in the form of gas than in the form of oil. The gaseous component of our recent discoveries of oil fields exceeds the liquid not only in volume, but in actual weight and, therefore, in the total energy made available to society.

This predominance of gas over oil in many of our important oil fields is significant economically. With modern technique natural gas has become the raw material, not only of essential products such as synthetic rubber, but even of gasoline itself. In short, natural gas has come now potentially to be oil. In the light of this development we should revise our estimates of oil reserves to include the oil-equivalent of the associated natural gas. Thus the figure we ordinarily quote as representing the estimated volume of our proved reserve of oil in the United States would be doubled; the present estimate of more than 19-billion barrels would become nearly 40-billion barrels.

Origin of Oil

We have come generally now to associate the origin of oil with a tropical environment. We think of the equable, humid climates of past geologic times when an incredibly prolific vegetation flourished throughout eons of time over the great, swamp-like lowlands of the earth's surface; we think of the myriad microscopic creatures teeming in the warm waters of the adjacent seas. This luxuriant former life captured the sunlight of one age after another and bequeathed it to us in our great stores of fossil fuels. The trees of the old forests left to us their carbon; through countless cycles of life and death it accumulated beneath the sour waters of the swamps, there to be buried and gradually transformed into coal. Similarly, the organic parts of marine creatures, buried in the muds of the sea floor, were converted at length into oil.

Most commonly accepted as probable source materials for oil in the earth are the organic, sea-floor oozes that become limestones and shales on compaction. Van der Gracht[1] believes that deposition of the richest source material takes place on the floors of seas either so saline (evaporites) or so foul with hydrogen sulphide (euxinic muds) that life is possible only close to the surface of the water. In these seas the bottom waters are devoid of oxygen, as well as of scavengers, and organic matter (such as plankton) falling to the bottom is neither devoured nor oxidized but is preserved to form oil. The dark euxinic muds from the floor of the Black Sea Contain 23% to 35% of organic matter against about 2.5%, average, for recent sediments, as reported by Parker Trask.[2]

Carbon can now be combined with hydrogen in the laboratory pretty much at will to form the whole series of hydrocarbons that occur naturally. S. C. Lind[3] of Minnesota University cites reactions, either thermal or ionic, by which progression both up and down the hydrocarbon series is possible, starting with any member. Whether we use as raw materials cellulose, or asphalt, or methane, all the combinations we find in oil can be synthetized by appropriate control of temperature, pressure, catalyst influence, alkalinity, and other factors.

The processes most frequently suggested for the conversion of carbon and hydrogen into oil within

---

[1] The Science of Petroleum, page 58, Oxford University Press, London
[2] Ibid., page 43.
[3] Ibid., page 39.

the earth's crust are bacterial action, alpha radiation and temperature. Bacteria are known to thrive in the uppermost layers of normal (not too saline, not too foul) sea-bottom sediments. Apparently they remove the nitrogen from organic muds, with the consequent enrichment of the potential raw material in the essential elements, hydrogen and carbon. In the presence of catalysts, alpha radiation, common in the earth's crust, has been shown by Lind [4] to transform hydrocarbons and to promote the union of hydrogen and carbon. Perhaps alpha radiation, then, combines hydrogen with carbon, converts organic matter to hydrocarbons and polymerizes one hydrocarbon to another in the earth where natural catalysts abound. If changes due to heat are involved in the natural synthesis of oil in the earth, they must proceed at relatively low levels. We require higher temperatures for distillation, cracking, hydrogenation, and polymerization in our laboratories than could have been tolerated by some of the constituents we find in natural oil. Within the earth's crust longer time may have compensated in some degree for lower temperature. Equally important may have been the presence of effective catalysts. The choice of catalysts in modern catalytic cracking has recently modified significantly our older thermal cracking practice.

The evolution of oil in the earth's crust through

[4] Loc. cit.

geologic time is marked by a decreasing molecule size and a concurrently increasing hydrogen-to-carbon ratio. Young oils are characteristically undersaturated; they are deficient in hydrogen and contain but little gas; they are asphaltic, made up of large or heavy molecules. Oil in geologically old rocks, on the other hand, is often paraffinic, with small, fully-saturated molecules, and is typically associated with large volumes of free gas. These facts suggest a progressive natural "cracking" in nature's laboratory, accompanied by parallel progressive "hydrogenation". Contamination of oils with sulphur or oxygen, or their compounds, arrests, or even reverses, this progressive cracking and hydrogenation. In consequence and as exceptions to the general rule heavy and unsaturated oils are occasionally found in geologically old rocks. These progressive changes (hydrogenation, cracking, polymerization, etc.,) require energy and therefore would not proceed spontaneously in oil in the earth, but the reactions generally involve so little chemical energy that Lind [1] believes the reaction once started proceeds easily in every direction through progressive steps.

Deserving more emphasis than most of us accord it is the fact that uniformly oil in the earth embraces an almost endless series of intimately related hydrocarbons in complex mixture. Was the parent source

[1] Loc. cit.

material so incredibly diverse in constitution? Or is it more reasonable to suppose that there is in oil pools an inherent tendency toward equilibrium which is only satisfied when each closely related member of this multitudinous family is finally present? If this inherent tendency really exists, then the most complex mixture of hydrocarbons might have evolved from a single simple source material. H. A. Wilson[1] has shown the theoretical existence of such an equilibrium, dependent on pressure and temperature, between each member of the hydrocarbon series and the next member both above and below; from methane at the light end through liquid to solid hydrocarbons at the other extreme.

To summarize, it may be said, in view of the evidence, that the synthesis of oil in the earth may be accomplished by much the same reactions we have discovered, independently of nature, and now use in our man-made manufacturing plants above ground. The rapid changes we bring about through extreme pressures and temperatures are accomplished more slowly and at lower temperatures but just as completely in earth's own laboratory underground.

Migration of Oil

A great oil pool includes so large a volume of hydrocarbons that the constituents must originally

[1] Proc. Royal Soc. A 120,247 (1928).

have been widely disseminated. The migration of disseminated hydrocarbons to form oil pools in the earth is probably contemporaneous in large part with the early stages of compaction of the source beds, when the material through which movement proceeds is still practically an ooze. Because they flow more readily through small openings gases would migrate farther than oil. Marine oozes, reacting to the heat generated by the internal friction of progressive compaction, might yield great volumes of hydrocarbon gases and, given access to continuous porous beds, these gases should move long distances freely. Migration is essentially a flow in the direction of lower pressure. Thus migration would be generally upward because only in that direction is ultimate relief from pressure to be obtained. However, because vertically upward movement may be prevented by impermeable overlying beds, lateral movement up-dip through gently tilted sands would commonly take place.

Traps, such as arches, domes or other convex recesses in the impervious roof of the porous layer, encountered by liquid and gaseous hydrocarbons migrating under these conditions, would gradually fill from the top downward. These traps, when completely filled, would spill around their base, just as air, entrapped under an inverted saucer in a pan of water, would escape around the edge as the saucer is tilted. Moreover, as accumulation proceeds under these conditions any liquid constituents are displaced from the top of the dome by the gas and, as the trap fills, the liquid is expelled first. In time the trap remains filled entirely with gas, the liquid having been displaced and having moved on up-dip until it encounters some other trap or escapes to the outcrop at the surface.

Under these conditions the traps near the trough of a sedimentary basin tend initially to contain a larger proportion of gas than the marginward traps. The latter would have captured some of the liquid oil forced out by the over-filling of the basinward traps as accumulation progressed. This relationship of a higher ratio of gas to oil in favorable structures progressively farther from the margins of the basin is exemplified in the distribution of some of the Gulf Coast oil fields. Even a trap which has filled entirely with gas comes in time to contain oil as some of the original gas changes to oil through the various processes, previously cited; hydrogenation, polymerization or ionization. Before equilibrium is attained the liquid oil may have come to exceed in volume the residual gas.

The Habitat of Oil in the Earth

Oil is found in the marine sedimentary rocks of the earth's crust, rocks that form from the beds of mud and sand and ooze spread out just off-shore on the floors of warm seas. If these areas of sedimentary rocks are extensive; if the depressions which they fill are deep; if they have persisted or been rejuvenated in one geologic era after another, with resulting over-lap and unconformity in the sedimentary sequence; if the whole series of sediments is thick then oil in the earth is abundant. But we must add a proviso: that movement subsequent to deposition has not too severely indurated, too intensely folded, or too widely disjointed the sedimentary beds. The habitat of oil in the earth, then, is the multitude of great marine-sediment-filled geosynclines, or basins, resulting from crustal downwarping.

The most common reservoir rock in oil fields is sandstone, but limestone, of course, also acts as host rock to accumulations of oil. Many of the world's greatest oil fields are in porous limestone or dolomite. Porosity in limestone may result from jointing, from irregularities along old erosion surfaces, or from leaching of layers made up in part of soluble minerals. In reef limestones pores are left by the loss of the original organic matter. Thick limestones are structurally competent and resist deformation that crushes and indurates associated shales and thin-bedded sands. Consequently, oil may be found at depth in or beneath heavy limestones in regions where less competent rocks nearer the surface have been metamorphosed beyond the point set by White's fixed-carbon index for the conversion of liquid hydrocarbons to gas.

Geologists have stressed the control exercised by geologic structure over the accumulation of oil. The recognition of this control marked the inception of systematic oil-finding. The theory of anticlinal accumulation of oil was elaborated by the earliest geologists who studied the problem, and oil-finding which previously had been largely fortuitous was immediately revolutionized. But this beautiful conception, perfectly valid in principle, has often actually led us astray in the practical search for oil. Geologists have relied on anticlinal structure of surface beds as a guide to oil in unexplored regions more than such evidence warrants. The first step in searching for oil in virgin territory is to drill a test on the most conspicuous anticline to be found. But adherence to this practice has often retarded discovery; the drilling of an unsuccessful first test on a conspicuous anticline tends to condemn the whole region.

Why do conspicuous anticlines so often fail us? We have observed that oil fields are delicate, balanced systems that do not often survive pronounced deformation of the containing rock layers. Deformation often becomes more intense at depth than it is at the surface; it increases with the thickness of the overlying section. Accordingly, where deformation has proceeded far enough to throw the beds at the surface of the earth into conspicuous folds it has frequently so broken and indurated the rocks at depth on the same line of folding as to have expelled from them any gas or oil they may once have contained.

In the early exploration of the Gulf Coast of Texas drilling was concentrated upon favorable structures that were conspicuous at the surface. Thirty years elapsed before we recognized the presence of and began to explore the deeply buried favorable structures that can hardly be discerned in the surface rocks. In the depths of the Gulf Coast basin even stratigraphic traps where almost no structural control is discernible have at length come to produce oil in relative abundance. The early production on the conspicuous structures now bulks small in proportion to that already obtained from these stratigraphic traps.

The history of exploration in the Maracaibo region of Venezuela repeats this procedure. Geologists mapped conspicuous anticlines around the margin of the basin. As in Texas, decades elapsed before the real oil fields were found far out in the structurally featureless central portion of the basin. Similarly, in the development of Eastern Venezuela early exploration was planned to test favorable surface structures, or the vicinity of gas and oil seepages, on the margins of the Orinoco Basin. In this case, exploration around the margin was successful in discovering oil fields, but years later a larger number of oil fields was found out in the middle of the basin.

As geologists, then, we should remember that anticlines are not indispensable to the accumulation of oil and that many of the important oil fields of the earth are not marked by conspicuous anticlinal structure in the overlying rocks at the surface.

To set down a few principles for our guidance in exploration:

(1) conspicuous anticlines near the margins of depositional basins have not generally proved to contain oil in large volume; where they do contain oil it is usually reasonable to believe that larger accumulations exist farther out in the basin;

(2) oil is found most frequently in proximity to marine organic shales or limestones;

(3) unconformities and overlaps of one series of beds on another provide traps for and control the accumulation of an impressive number of large oil fields;

(4) so many of the large oil fields of the earth have been discovered in relatively undeformed beds, under thousands of feet of cover, in the central portions of large depositional basins, far from their margins, as to suggest that such an environment may well be the preferred habitat of oil in the earth.

II

WHERE OIL IS

II

WE HAVE looked briefly at the environment of that remarkable family of hydrocarbons — gaseous, liquid and solid — oil in the earth. We found certain attributes of this environment to be so characteristic that their absence in any accumulation of oil large enough to be important commercially marks that accumulation as exceptional. Reviewing these attributes briefly we reach two general conclusions pertinent to the search for oil in the earth:

(1) oil is in the porous, marine rocks of the earth's crust, in proximity to shales or limestones rich in organic matter, where it is trapped, imprisoned between impervious beds above and the heavier brines (fossil sea water), which also permeate these rocks, below.

(2) oil in the earth is transient. The delicate equilibrium of a natural oil reservoir with its balanced series of liquid and gaseous hydrocarbons, under extreme pressures, cannot withstand violent deformation of the enclosing rocks, nor survive even moderate deformation over geologically long periods.

From these facts it follows that, in an earth prone to the periodic travail of crustal deformation, oil will be found most commonly in the geologically young regions where the pores of marine rocks have not been completely destroyed by induration; where the lapse of time has not yet sufficed to complete the overturning of the physical and chemical equilibria that constitute oil pools. It is geologically young rocks, the Cenozoic (Eocene, Oligocene, Miocene and Pliocene, in order of decreasing age) that contain most of our oil reserves; and next in rank is the Mesozoic, or "middle aged" series of rocks (Triassic, Jurassic and Cretaceous). The Paleozoic rocks (Cambrian, Ordovician, Silurian, Devonian, Mississippian, Pennsylvanian, and Permian), the old rocks of the earth's crust, retain the oil that concentrated itself within their porous members during the early stages of compaction, only where an unusual combination of circumstances has warded off, through one geologic revolution after another, the induration that is the ultimate fate of clastic rocks.

Kansans may register skepticism at the generalization that oil is most often found in the young rocks of the earth. With their own prolific oil fields in Paleozoic rocks they will naturally stress the abundance of oil in the Paleozoic. They will even recall that the world's oil industry was born in the Paleozoic rocks of Pennsylvania. Nevertheless, despite their experience at home, Americans must realize that over the earth as a whole oil is far more abundant in the young than in the old parts of the earth's crust.

True, much of the past production of oil within the United States has been obtained from Paleozoic rocks; out of a total production of some 24 billion barrels prior to 1942 about 11.5 billion or nearly 50 per cent has come from the old rocks. But of the 19 billion barrels of proved reserves in the United States (proved reserves being the oil already discovered and developed in contrast to other possible resources of oil, the presence of which may be suspected but is not proved), only about 5 billion barrels, or 26 per cent, is in the old rocks; and in the world outside the United States, with a past production of approximately 14 billion barrels and proved reserves of 24 billion barrels, only about 1 per cent of either past production or reserves is older than the Mesozoic. In summary then, out of total past discoveries of oil in the earth amounting to 81 billion barrels, only 17 billion barrels, or about 20 per cent, has been found in the old rocks.

We have learned that oil is a normal constituent of the marine sedimentary rocks all over the earth. We know that marine rocks make up some 40 per cent of the total land area of the globe, giving us an area of some 25 million square miles within which we can reasonably expect to find oil. But for the most part the earth's store of oil still remains hidden from us. We have found only the most conspicuous accumulations of oil. When we peer into this obscurity surrounding the occurrence of oil, two regions emerge as the principal reservoirs of oil in the earth. Since these regions lie close to the opposite ends of an earth diameter they might be called the "petroleum poles" of the earth. Or, in keeping with what Carl Becker would designate as the "climate of opinion" prevailing in the world today, these regions might be described as marking the earth's "oil axis". One end of this axis is the Near East, the environs of the Black, Caspian, and Red seas, the Persian Gulf, and the eastern end of the Mediterranean Sea in the Old World. Within this region lie the major oil fields of Russia, Iran, Iraq, Arabia, and Egypt. The other end of this oil axis is the land margin of what might be called the mediterranean region of the New World, that geologically modern crustal depression now occupied by the Gulf of Mexico and the Caribbean Sea; here lie the great oil fields of Venezuela, Colombia, Mexico, and of the Gulf Coast of the United States. Of the proved oil reserve of the United States, more than 9 billion barrels, or nearly one-half of the nation's total, lie within this region tributary to the Gulf of Mexico. Altogether, nearly 30 billion barrels of oil, or about two-thirds of the total proved reserve of oil in the earth, is contained in these two regions at the opposite ends of the earth's oil axis. Just as the modern ice cap at the South Pole exceeds that at the North Pole, so the Old World end of the oil axis, with some 16 billion barrels proved reserves, dominates the New World end and thus becomes the outstanding oil reserve in the earth.

Unindurated marine rocks, as we have learned, are the most common host rocks to oil in the earth. The ends of the oil axis are what they are — sites of the greatest natural reservoirs of oil in the earth — precisely because they fulfill to a superlative degree these essential conditions of accumulation of oil in the earth's crust; they mark the sites of the greatest segments of the earth's crust composed of only moderately deformed, geologically young, marine rocks. From Mesosoic forward through the Cenozoic times marine basins in each of these regions have been loaded with sediment from surrounding land masses. At times in both regions extensive desiccating seas have prevailed, showering their bottoms with evaporites rich in organic remains. At times euxinic muds also rich with the remains of former marine life have spread in abundance over these basins. Crustal warping in these regions since the close of the Mesozoic has compacted the sediments and tilted the beds into attitudes admirably suited to arrest and trap moving hydrocarbons, but has not been severe enough to destroy the resulting accumulations of oil. So were established the conditions under which oil fields evolve.

In Iraq and Iran, at the Old World end of the

axis, the principal producing horizon, the Asmari dolomite of Eocene age, belongs to an evaporite series and is closely associated with salt and anhydrite. In Russia, farther north, younger beds, shales and sands, rather than limestone, are the principal source of oil, whereas to the south the oil fields recently developed by American groups in Arabia are in sandstone older in age than the Asmari limestone. This region, the Near East, is the best hunting-ground for oil on earth. This fact has long been apparent; numerous prolific oil fields are already under development but exploration is only well begun and much greater reserves of oil undoubtedly remain to be discovered. Much of the promising territory lies within the boundaries of the Union of Soviet Republics and is, therefore, closed to exploration by the outside world. But excluding the territory of the Soviets, the neighboring countries to the south, Iran, Iraq, Arabia, Afghanistan, Syria, Palestine, Turkey and Egypt, still promise more oil from future discoveries than any other part of the earth.

The oil-bearing rocks in the Gulf of Mexico-Caribbean Sea region at the other end of the oil axis range in age from early Mesozoic through the Cenozoic, and include dolomitic limestones in a sequence of evaporites, as well as sandstones intercalated with ordinary marine sediments.

The Orinoco Basin, in eastern Venezuela, is filled with Cenozoic sediments. Overlapping in turn on the old crystallines of the continental shield to the south, these rocks dip northward over a distance of 150 miles to abut against an east-west mountain range of Mesozoic rocks marking the northern coast of Venezuela. Oil is found in Eocene, Oligocene, Miocene, and even Pliocene sands, and along unconformities which separate these several series. A similar, unexplored, but even larger basin occupies the region of the head waters of the Amazon River, extending for hundreds of miles along the Andes Mountain front. This remote tropical empire undoubtedly houses tremendous hidden stores of oil, resources which could be developed commercially at costs only slightly higher than those currently prevailing.

The Lake Maracaibo fields in western Venezuela constitute one of the greatest developed sources of oil in the Western Hemisphere. Venezuela in recent years has surpassed all other nations except the United States and Russia in the volume of her oil production, and most of this production has come from Lake Maracaibo. The basin surrounding the lake lies in the angle between the central and eastern ranges of the Andes Mountains immediately north of their juncture. This basin has been described as the richest source of oil in the Western Hemisphere. Much exploration still remains to be carried out in this area.

In Colombia, in Mexico, and in our own Gulf

Coast region are other prolific fields; nowhere have the possibilities been exhausted. The valley of the Magdalena River in northern Colombia, an immense lowland lying, like Lake Maracaibo in Venezuela, between two main prongs of the Andes, is already a large producer of oil. The pampas of southeastern Colombia, a part of the great province along the eastern front of the Andes, is still to be explored. Mexico, once second in rank among the oil-producing nations of the world, in only temporary eclipse as a result of governmental restrictions, possesses magnificent potential oil resources. In a different climate of opinion Mexico will certainly be again the objective of sustained exploration and will again assume her former importance as an oil-producer. Central America, between Colombia and Mexico, and the islands of the West Indies group, have been explored only very cursorily. These regions should reward careful search with commercial oil fields.

Of lesser importance than either end of the earth's oil axis are the fringing areas of Mesozoic and younger rocks common to both shores of the Pacific Ocean. Just as the Pacific is girdled with a "ring of fire" in its encircling belt of volcanoes, so the continental shelves on either side of the Pacific basin are veneered with geologically young, marine rocks, oil-bearing where they have escaped severe deformation. From this province California has developed its copious present flow and its large reserve of oil. From this province, also, Peru and Ecuador draw their production. Alaska, although not yet an oil producer, includes a part of the Pacific's border of marine rocks and may well yield oil in the future. In the opposite direction the extreme tip of Chile, all but touching Antarctica, also promises to produce oil. On the other side of the Pacific the large islands of the Malayan Archipelago, Borneo, Sumatra, Java, and New Guinea, produce currently and promise to produce in the future enough oil to rank them ultimately along with California as one of the earth's major oil sources. Farther north in the Orient, Japan and Russia have exploited smaller oil accumulations on the islands of Honshu and Sakhalin.

Within the United States our outstanding oil province is the Gulf Coast region, varying up to 300 miles in width, stretching over a length of 1200 miles from Florida to the Rio Grande and already highly productive for more than half its length. Most of our past production has come from this province; and far into the future most of our exploratory activities will be devoted to it and most of our oil will continue to come from it. California, with a long record of copious production, is next in importance. The Mid-continent region, northern Texas, Oklahoma, Kansas and Nebraska; the Central Eastern states, Illinois, Kentucky, Indiana and

Michigan; and the old Appalachian fields in Pennsylvania, New York and West Virginia have furnished a large proportion of America's oil in the past, and in these states important new oil fields will be found despite the thorough exploration already accorded them. And even more promising for future discoveries are two other regions: the Permian Basin of West Texas and the numerous intermontane basins of the Rocky Mountain states.

The Permian Basin of West Texas (south of the Panhandle) and southeastern New Mexico covers an area of 50,000 square miles. Throughout this area the surface is underlain to depths of several thousand feet with rocks formed from the oozes, muds and sands spread out over the floors of rapidly evaporating seas during the arid climates of Permian times. These rocks are called evaporites because in large part they are chemical precipitates from concentrated waters; they include limestone, dolomite, anhydrite, gypsum and rock salt, each of which is present in aggregate thickness of thousands of feet. The Permian Basin has been under development for the last twenty years and has already given up large volumes of oil.

But this part of the earth's surface was a basin of deposition long before the Permian seas spread over it. Throughout Paleozoic time organic muds and oozes were being deposited on the bottoms of seas in this region and the rocks into which these sediments were transformed have also become likely sources of oil. The exploration of these older rocks, now buried beneath the whole Permian series, is only well begun. Oil has already been produced from them at various places through wells that go to depths of 10,000 feet, or more. In spite of the great depth these wells yield oil in such large volume as to be very profitable. The principal reservoir bed so far developed in the older rocks is a dolomitic limestone of Ordovician age, near the base of the Paleozoic column. It is already apparent that tremendous reserves of oil are to be discovered in this prolific horizon.

The intermontane basins of the Rocky Mountain states, Wyoming and Montana, in particular, have already been producing oil for several decades. Together with the profoundly down-warped portion of the Williston Basin immediately in front of the Rockies, these basins constitute a potential source of much more oil in the future. Most of the discoveries so far made in them are in Mesozoic rocks near their margins, but Paleozoic rocks underlie the Mesozoic and this lower, older series has also been proved to be oil-bearing. The central, deeper parts of these basins, as yet all but unexplored, offer great promise.

To students in geology at K. U. the Williston Basin offers special interest. Close at hand, the largest basin of deposition on the North American continent, it remains far less thoroughly explored than many smaller, less promising basins. Moreover, it is filled with the same formations that contain the great accumulations of oil in Kansas. It spreads over parts of the states of Nebraska, North and South Dakota, Wyoming and Montana, and continues several hundred miles farther northward into the provinces of Saskatchewan and Alberta in Canada. There it joins the Mackenzie River Basin which extends on northward to the Arctic Coast. The floor of the Williston Basin tilts gently to the southwest from the older crystallines of the Laurentian Shield in central Canada to rise again abruptly out of a deep trough along the Rocky Mountain front. This simple basin structure is modified by a discontinuous, almost entirely buried ridge of the pre-Paleozoic rocks jutting out from its southern margin. This barrier is marked by the Black Hills in South Dakota, and by the Baker-Glendive flexure farther north-northwestward in Montana. The basin to the west of this barrier is often considered separately and is designated as the Powder River Basin.

Within the Williston Basin are widespread evidences of oil. In the Powder River Basin, along the Rocky Mountain front, Paleozoic and Mesozoic rocks alike have already proved productive; even the Eocene rocks, although they are of continental origin at the surface, may prove at depth to take on the character of host rocks for oil.

Challenging to the oil man also is the occurrence of the Athabaska Tar Sands, spreading over thousands of square miles in the region of the Athabaska River in the extreme northern portion of the Williston Basin. These tar sands display a volume of oil greater than any other so far discovered on earth. Estimates place the oil-equivalent of the Athabaska Tar Sands at from 100 to 250 billion barrels; yet the total discoveries of all time over all of the earth, as we have already noted, do not exceed 81 billion barrels. The Athabaska Tar Sands are not a variety of oil shale as has been commonly assumed; they are literally oil-saturated sands. Oil shales contain kerogen a compound which, like coal, can be converted into oil through retorting under appropriate pressure-temperature conditions. But the Athabaska Tar Sands contain oil itself. At the surface the oil is waxy and inspissated, but underground the tar sands yield a heavy oil. There is little doubt that as a local source of oil they could be exploited commercially today: eventually they may well come to be looked upon as one of the important energy resources of North America.

Let us continue our northward journey even farther, up to the very coasts of the Arctic. Our conception of the environment required for the generation of the source materials for oil, sunlight,

warmth and life in profusion, is not easily recon-
cilable with the icy, barren wastes of the Arctic.
Our difficulty is at once resolved, however, if we re-
mind ourselves that the earth's polar ice caps, which
we unconsciously think of as permanent, are in
reality veritable newcomers, geologically. Throughout the greater part of geologic time life has abound-
ed in the polar regions. Fossil plants in the rocks
of the Island of Spitzbergen prove that well up into
the Cenozoic Era tropical palms and ferns flourished in latitude 80 degrees north, where today only
the hardiest shrubs survive and even the mightiest
trees attain a height of only two inches.

In the past, then, geological conditions in the Far
North have been favorable for the genesis of oil.
The waters that cover the North Pole occupy a
depression in the earth's surface which, although
it may never have been of extreme depth, must have
persisted throughout most of geologic time. These
waters are properly to be thought of as a sea rather
than as an ocean. Many of our maps portray the
"Arctic Ocean", but the Arctic waters are in fact
land-locked and are, therefore, a sea. Like the Caspian, Black and Mediterranean seas and the Gulf of
Mexico-Caribbean Sea depression, the Arctic mediterranean sea has been loaded with sediments from
adjacent land areas through one geologic period after another. Like these other two regions of land-
locked seas, which, as we have noted, are the sites

of the earth's principal oil resources, the Arctic is an area of extreme promise for oil.

And if we look for oil there even the most cursory search reveals its presence. Near Point Barrow, on the Arctic coast, in northernmost Alaska, in an extensive area of Mesozoic rocks, are some remarkable seepages of oil:[1]

"The main petroleum flow moves southward down the slope for 600 or 700 feet to a lake. This active channel is 6 to 10 feet wide . . . . "

No one can doubt that such evidence justifies exploration for oil fields.

Canada has long had a producing oil field at Fort Norman, 65 degrees north latitude. A small refinery at Fort Norman has manufactured gasoline and fuel oil for local use for many years. Recently, under the spur of war, exploration has been renewed and a great oil field has been developed. This locality is in the central portion of the Mackenzie River Basin, which extends over a large area from the northern limit of the Williston Basin to the Arctic Sea. Wholly unexplored except for the Fort Norman field, it is very promising for oil.

V. Stefensson,[2] the celebrated Arctic explorer, reports live seepages of oil on northern Melville Island, 500 miles north of the Arctic Circle. The

[1] "A Reconnaisance of the Point Barrow Region, Alaska". Sidney Paige, W. T. Foran and J. Gilully. U.S.G.S. Bull. 772 (1925) page 23.

[2] Personal communication to the author.

rocks in which these seepages occur can be traced northeastward to the extreme northern coast of Greenland where their character still remains favorable for oil.

Directing our attention still farther eastward over the lands that fringe the Arctic, we learn that the Soviet geologists have reported oil both from natural seepages and from test wells, along the northern coast of Siberia. These evidences of oil characterize an area 3000 miles in length. Elsewhere in Siberia, as well as on the western flank of the northern Ural Mountains, oil has been found in commercially important quantities.

Altogether there are 1.5 million square miles favorable for oil north of the sixtieth parallel of north latitude; one-third of this area lies in the Western Hemisphere. By way of comparison, the United States contains 2.4 million square miles of favorable territory. Civilization has been defined as the record of the slow but persistent movement of humanity northward along a circular front at the center of which lies the North Pole. However this may be, these apparently great potential oil resources in the Far North may mean much to the future since the most direct routes of air transportation from our own great cities to the Orient and to northern Europe traverse the Arctic.

The oil resources of Russia merit special attention. Still largely undeveloped, still, indeed, largely unexplored, they are nevertheless without question potentially far greater than those of any other nation. They include not only a considerable slice of the Near East province; they include also adjacent oil fields at intervals throughout a zone that continues northward from the Caspian Sea development, along the western flank of the Ural Mountains to the Arctic coast, a distance of 2000 miles. East of the northern end of the Urals are the occurrences already referred to along the Arctic coast of Siberia. Eastward from the Caspian Sea oil-bearing rocks are encountered as far as the borders of China, nearly 1500 miles distant. And 1500 miles farther east, near Lake Baikal in southeastern Siberia, new oil fields are under development by Soviet engineers. There still remain to be mentioned the extensive development work recently undertaken in oil-bearing rocks in the Yakutsk region on the upper Lena River, 1000 miles northeast of Lake Baikal, and the older oil fields on Sakhalin Island and Kamchatka Peninsula on the Pacific Coast.

We have sketched the most prolific sources of oil in the earth. Subordinate occurrences, some of which promise important future production, include the Mesozoic and Paleozoic rocks along the eastern front of the Andes in Bolivia, Paraguay and Argentine. Poland, Roumania, Germany, Hungary, France and England each produces moderate volumes of oil. Burma, fringing the Indian Ocean on the southern edge of the Asiatic continent, with its impressive record of past production, together with India, promises rich rewards to further search. Other parts of the earth, not now productive, doubtless will come to yield oil in the future. Over much of the earth exploration has hardly begun. Africa, for example, is all but virgin territory.

Remarkable for an almost complete absence of oil is the continent of Australia. Except for apparently small quantities of rich "kerosene shales" and a little natural gas, no evidences of oil have been noted. Neither has New Zealand developed oil of commercial importance. Nevertheless, it is hardly to be doubted that persistent exploration in either Australia or New Zealand would be rewarded with discoveries of oil.

Of approximately 81 billion barrels total production to come ultimately from all past discoveries of oil in the earth, the United States alone accounts for nearly 44 billion barrels, or 54 per cent of the total. Yet the land area of the United States, 3 million square miles, is only 5 per cent of the total land area of the earth (60 million square miles). If oil were uniformly distributed and exploration were uniformly engaged in over the earth other countries should eventually account for 95 per cent of the total discoveries instead of 46 per cent as at present.

In an unpublished study of oil-field discoveries

over the earth which I have been permitted to consult, Eugene Stebinger and L. G. Weeks have made a comparative analysis of oil-finding experience. Excluding areas of igneous and metamorphic rocks as impossible for the production of oil, and comparing only sedimentary areas in which oil might reasonably be expected, the United States (not including Alaska), with 2.4 million square miles, accounts for 11 per cent of the earth's total sedimentary area (22 million square miles). Of the land area of the United States 80 per cent is composed of sedimentary rocks likely to contain oil, whereas for the earth, excluding the United States, only 34 per cent of the land area is of this character. Even so, if sedimentary rocks yielded oil uniformly over the earth, the United States might expect ultimately to produce only 11 per cent of the earth's total instead of 54 per cent, its contribution to date.

Stebinger and Weeks have classified sedimentary areas still further as to their promise for oil, retaining for consideration only the more favorable areas where marine sediments occur in thick section over extensive basins. Of these favorable areas the United States is found to have about 0.9 million square miles or 15 per cent of the known total area of 6 million square miles for the earth. This classification makes 38 per cent of the sedimentary area of the United States favorable for oil; whereas, excluding the United States only 26 per cent of the earth's sedimentary area is favorable for oil. If the promising sedimentary rocks prove to be equally productive over the earth, the United States will ultimately produce only about 15 per cent of the earth's oil.

Of coal, which is also found only in sedimentary rocks, the United States possesses an estimated 3420 billion tons, or 45 per cent of the earth's total of 7530 billion tons. If 45 per cent of the earth's coal is to come from the 11 per cent of the total sedimentary area included with in the United States, perhaps 54 per cent of the earth's oil might also come from the same small sedimentary area. Nevertheless, the participation of the United States in the earth's total oil when the account is finally cast up, seems likely to be smaller than its share in that which has already been discovered. When we inquire into the record of oil-finding we learn that search for oil has been far more intensive in the United States than abroad. Much oil remains to be found in this country, but far greater volumes should come from future discoveries in other countries.

Some factors in the problem of where oil is in the earth, however, elude graphs and statistical calculation. These factors are not resolved by any study of the record of past discoveries. They bear no discernible relationship to the physical character of the earth's oil reservoirs. If, for example, the richly oil-bearing region we call Kansas today had slipped from her present moorings at the inception of our Civil War, say, and surreptitiously traded places with an equal area somewhere in Europe, translating herself bodily to become part of the national territory of any one of a half dozen Old World powers, the chances are that all of her oil fields would have remained still undiscovered. By the same token, whatever region in Europe had come over here to take the place of Kansas would probably have been gushing oil today, just as Kansas is.

The oil fields of Kansas and other western states were developed by a curious force like nothing ever recorded in any other country. This force was born with Oil Creek, centuries after the Old World had become aware of oil in the earth, but before anyone there had really made use of his knowledge. It grew and developed with the spread of oil through Pennsylvania, and gradually pushed westward, leaving a train of oil fields in its wake. The states across which it travelled yielded up oil, more or less without regard to their geological constitution. To be sure, great areas of metamorphics and crystallines like the iron ranges of Minnesota and the Ozarks in Missouri remained obdurate, but other regions responded one after another to the call of the "wildcatter;"[1] states as diverse geologically as Kentucky and Ohio, Illinois and Texas, Kansas and Louisiana, each developed great oil fields. Continental France today possesses no large oil field, yet if France had lain across the path of the American oil-finder I have little doubt that France too would have taken her place in this procession of oil-producing regions. It is the genius of a people that determines how much oil shall be reduced to possession; the presence of oil in the earth itself is not enough.

Men who grew up among the oil fields of Pennsylvania or West Virginia came west and, knowing nothing of geology—indeed almost in defiance of geology—drilled wells that discovered new oil fields. A little ahead of these pioneers the Fortieth Parallel Survey traversed this same westward course in the early 1870's, with intellectual giants like Clarence King and Frank Emmons searching with trained eyes for possible evidences of mineral deposits. Yet this classic of geologic exploration reported nothing to justify the belief that myriads of oil derricks would soon come to dot the landscape. Where, for example, could a less promising location for an oil test have been found than the plains of western Oklahoma, where Ponca City now stands, when the prospecting that resulted in the discovery of that

[1] A "wildcatter" is the driller of a "wildcat" well; that is an exploratory well drilled in search for oil or gas in virgin, unexplored territory outside the proved boundaries of any producing oil field.

great oil field was undertaken? Surely, these rolling prairies bore no physical resemblance to the cuestas and hog-backs of Pennsylvania in the Appalachian foothills where American pioneers had learned to drill for oil. What then guided these adventurers in their sensational discoveries of oil fields across the entire width of a continent?

Reflection upon this record and similar experiences elsewhere arouses a suspicion that physical conditions in the earth's crust impose fewer and less formidable obstacles to the development of commercial oil fields over the earth than do some of our mental and social habits. If American experience is any criterion, oil in the earth is abundant. When we search diligently for it, we find it. Over most of the earth this diligent search has been wanting. Until the earth at large has been explored as the United States has been explored, oil in the earth, no matter how plentiful it may be, will remain undiscovered. National temperament may constitute a barrier to oil-finding just as baffling as the most complicated geologic pattern. Whatever the geological conditions may be and whatever technique we employ, we find oil in the earth very rarely unless we have first acquired an appropriate mental attitude. The very state of mind of the social order as well as of the individual is involved in successful oil-finding. Brazil and China have no oil today but they might well have oil had their nationals searched for it in the American fashion. Gold is where you find it, according to an old adage, but, judging from the record of our experience, oil must be sought first of all in our minds.

Where oil really is, then, in the final analysis, is in our own heads!

III

WHO FINDS OIL AND HOW

III

OF ALL the oil so far discovered and developed in the earth, some 81 billion barrels or 54 per cent of the total, is accounted for by the search within the United States (excluding Alaska) alone. This country constitutes only 5 per cent of the total land area of the earth, contains only 11 per cent of the total area of marine rocks on the earth, and only 15 per cent of the earth's total area of important, marine-rock filled sedimentary basins, of a character favorable for the occurrence of oil fields; yet the United States has furnished 54 per cent of all the oil discovered to date. Per unit of total area, therefore, we have found more than 20 times as much oil as has been found in the rest of the world and per unit of area favorable for oil we have found in our own land 7 times as much oil as the rest of the world has found. Should this discrepancy be interpreted to mean that the average square mile of our country has been endowed by nature with 7 times as much oil as the average square mile of other countries? Or does it mean that we have sought oil in the earth more effectively than have other nations?

There is little evidence to support a conclusion that America has been 7 times more richly blessed with oil than other countries. But there is abundant evidence that Americans search for oil more assiduously and find oil more effectively than any other people on earth. In fact it is Americans who find the oil in the earth. Not only do Americans find oil in their own country but it is Americans who have found much of the oil in other countries. Even where foreign capital has financed oil-finding enterprises the actual work of exploration has commonly been performed by American technologists, geologists and engineers. The oil-fields of Mexico, Colombia, and Venezuela are outstanding achievements of Americans in oil-finding in the Western Hemisphere. All over the world, American technical skill as well as American machinery has borne the brunt of the exploratory effort. Americans were responsible for the first discovery of oil and the completion (much later) of the first commercially productive well in England. Recently, Americans found the first oil to be discovered in Hungary and develped a prolific field in that country, long starved for oil. The only large oil field so far found in Germany, where exploration has been methodical and oil has been produced in small quantities for generations, was discovered by Americans.

Oil-finding and the development of oil fields, then, are largely American achievements. Americans have stamped the modern industry with their character, their methods and their machinery. How has it come about that Americans have found more oil and developed oil resources more rapidly and more extensively than other peoples? It is worth while to seek an answer to this query because in very large part America's high standard of living is sustained by the abundance of cheap energy she derives from her bountiful oil supplies.

Three factors combine to make Americans stand out as discoverers and developers of oil in the earth up to the present. These factors are:

(1) A distinctive flair for effective team work between science and industry which arises out of the American concept of the dignity of labor.

(2) The freedom from political and social barriers to widespread exploration for oil beneath the surface of the earth in the United States.

(3) The adventurous, chance-taking spirit of the pioneer which pervades America and has impelled Americans to drill thousands of wells every year in search for oil.

In essence, finding and developing oil fields are geological enterprises. They involve the application of an inexact, immature science to a very precise, very complicated industrial activity. The research of the scientist, his mental habit and accumulated knowledge, must somehow be instilled into the behavior of the common laborer so effectively that through the mind and hand of the workman science guides and controls, accurately and without friction, the ponderous machinery with which, if we are to find oil, we must carefully probe down, mile after mile, into the crust of the earth.

Obviously here is a task not easy of accomplishment; but other formidable barriers obstruct the way to the conquest of a nation's oil resources. If only learned scientists, tractable workmen and heavy machinery were required, Americans would hardly excel in this field. American geologists are not superior to, if indeed, they are as good as the geologists of other countries. Workers and machines abound in many countries.

The prime requisite to success in oil-finding is freedom to explore: and only slightly less imperative is freedom to develop and produce the oil once it is found. In these freedoms (as in many other freedoms) America does excel. In the United States every farmer and every rancher owns and can freely dispose of the oil under his land. Nowhere else on earth are oil resources so readily accessible to the itinerant "wildcatter" as in the United States. Generations ahead of the rest of the world Americans developed their oil resources. Unless they had been free to drill exploratory wells at will they could not have accomplished this feat. The American oil industry could hardly have established itself in any European country. Only where the drilling of an oil well can be arranged for as simply as a citizen can make a contract with his neighbor is an oil industry able to function as the American oil industry has functioned. If the oil resources of a nation are held by the crown, or by the commonwealth, or if they are locked up inviolate under the estates of a landed aristocracy to which the common man is denied access, then no enterprise like the American oil industry can attach to them.

That science and industry are in harness; that social and political environments are favorable, are still not enough to account for our success in oil-finding. Indispensable to the success of the enterprise is the spirit of the pioneer, restless, daring, risking his all on any reasonable prospect of recovering more. Deep in the earth beneath us we discerned a new frontier. As pioneers we set out to explore this geological frontier with the same ingenuity and resourcefulness that marked our forefathers' conquest of our geographical frontiers. This vertical frontier exists for every nation but no other nation has ever explored its depths as we have done. Yet the reward to us has been great—more oil than all the rest of the earth has produced and, in consequence, a standard of living higher than that of any other people.

Finding oil in the earth did not become important until after the automobile came into our lives and the first World War quickened the pace of naval performance. Then nations began to need oil. The larger American oil companies turned to geology (a little doubtfully) as a possible aid to their wild-catting ventures. Threatened scarcity stimulated them to build up elaborate geological departments. They offered employment to the graduates in geology from colleges and universities all over the country. The schools in turn rapidly built up courses of instruction in petroleum geology. Young men everywhere rushed to enroll in them, lured by the fascination of tracking down and reducing to possession this elusive quarry, the deeply buried oil field. Very soon the available supply of geologists exceeded the needs of the industry for purely geologic staffs. Unable to secure employment in geological departments which so far had not been too successful in their function, geologists began to accept work in other departments. With complete adaptability they got jobs as rough necks and roustabouts; they rose from these lowest ranks to become drillers and foremen; they served as scouts and lease men. With this turn geology, heretofore merely a narrow path to a specialist's job in oil-finding, widened into a broad avenue of approach to the whole field of oil production. In due time geologists, having acquired the necessary experience, financed and drilled wild-cat wells for themselves and formed their own producing companies. Other geologists, having started in various capacities with the larger producing units, advanced in rank until many of them came to hold responsible executive positions.

So it was that geologic principles gradually came to permeate the whole industry. For the first time, perhaps, the scientific method made itself felt in American industry. Analysis and persistent search for facts slowly replaced hunch, tradition and dogma; objectivity established itself over an ever expanding area. Coincident with this change, geology correspondingly increased its effectiveness in its own peculiar domain. As long as geology had been dispensed by a detached chief geologist, carefully insulated from any contact with other producing functions, only a mediocre success in oil-finding was attained. But as men with a geologic point of view began to take over duties and assume responsibilities outside their own limited activity and, finally, to influence decisions in the management of the enterprise, success in oil-finding became phenomenal.

An oil-field scout trained in geology was able to interpret the rumors, reports and observations coming to his notice with an understanding and perspective unattainable by his predecessor to whom geology had been a closed book. Geologists who themselves actually drilled wells had a genuinely first-hand knowledge of the formations encountered. Such wells were completed with a full comprehension of the stratigraphy and structure of the natural reservoir from which they were to drain oil. The lease man, with a first-hand knowledge of the geology of the area under negotiation, could hardly fail to deal more intelligently in the acquisition of leaseholds than the lease man armed only with a flair for trading. And, finally, the geologist in an executive position found an almost unlimited field over which he exercised advantageously his geologic acumen. Under the stimulus of this flow of geologic competence from every angle into the stream of industrial effort the tide of oil-field discovery mounted higher and higher. Dry hole averages decreased, proved reserves mounted, recoveries per well and per acre improved and unit costs declined all through the operation.

So it is that geology has steadily seeped out from the almost hermetically sealed compartment of pure science to which it was once confined, until now it saturates the entire structure of the oil-producing industry. This conquest of the industry by geology has been not unlike the process of metasomatism, to borrow a somewhat imposing term from the hard-rock geologists. Metasomatism is that important process in ore deposition whereby the invading solution, although it leaves the outward form or body of the host rock unchanged, nevertheless entirely transforms its intrinsic character, replacing the original internal constituents, molecule by molecule, with substitutes of its own selection. So has geology reformed the enterprise of producing oil.

The industry has re-fashioned its whole technique, assembling it anew in a completely geologic setting. Today no engineer determines an optimum rate of production, no drilling crew sets an oil string, no conservation officer promulgates a spacing order for a new field, but he draws importantly, albeit unconsciously, perhaps, on the collective geologic experience of the industry.

The modern oil-producing industry exemplifies a way of doing things that is peculiarly American. Only in a society where the man of culture, the so-called gentleman, labors freely with his hands, unembarrassed even at menial tasks, can a co-operative effort of this character attain full success. Regimentation and class distinction are alike fatal to it: the concept of dignity in labor is vital to it. In America beyond all other nations these free associations and this attitude toward manual labor have prevailed. By no mere coincidence, then, have most of the oil fields of the world been found by our own citizens. Search for oil fields succeeds best through the method of consummate team work which Americans have perfected.

Consider the alternative method which places geologists in one category, drilling experts in another, production engineers in another, etc., etc. Each group is insulated rigidly from the other; each reports to and is directed by a different authority. The geologist in his work is denied the great advan-

tage of that identity of viewpoint with the driller
and the production engineer on their problems,
which first-hand personal experience in these capacities would give him. The driller and the production
engineer, in turn, suffer from a similar lack of any
geologic understanding. Under these conditions
even the most brilliant individual performance is
handicapped. Inborn class distinctions do these
things to groups of fellow workers. There is less
class distinction and more true equality in America
and particularly in the oil-producing industry of
America than anywhere else on earth.

Although for a generation now the strategy of
American oil finders has been consistent in making
their venture a geological enterprise, their tactics
have changed repeatedly as one campaign merged
into another. But always in their search for oil
Americans have had the courage to drill exploratory
wells; many wells and deep. An accurate record of
the rocks encountered in the drilling of a deep well,
whether the well proves to be a dry hole or a producer is one of the most useful tools the geologist
possesses. Such records, or "well logs", can be obtained only by drilling wells. Well logs have improved as exploration has advanced; the driller's
pencilled record of what he infers from the behavior
of his "tools" is now supplemented by samples of
the rocks actually penetrated by the drill. An even
more valuable improvement is the "electric well

log"; a comparative analysis in the form of a graph of the varying electrical properties of each successive rock layer encountered by the drill. In the light they throw on the proper correlation of the strata penetrated by the drill the present-day electric well logs are to the old-time drillers' log as day is to night.

From the tens of thousands of wells drilled in America each year is revealed more of the geological character of North America underground than is known of any other continent. Some of these wells reward the industrialist by producing oil, but many of them produce none. As to these failures, the capitalist suffers a total loss for the moment, but every one of these wells, whether it produces oil or not, becomes a new and vastly exciting experiment to the geologist who thus delves into natures' own laboratory and freely reaps each year his vast crop of scientific facts. When these successive harvests of facts have been winnowed, their significance is brought to bear once more on the problem of finding oil. So in the end industry is compensated for its dry holes.

Exploration for oil in the United States is carried on continuously by thousands of independent enterprises which drill year in and year out twenty or thirty thousand test wells. Inspired by the financial reward that comes with success, they punch down their wells in any unexplored locality, regardless of whether the experts think well or badly of its promise. If the sand which constitutes the first objective fails to produce oil, the well goes on down anyway just on the chance there may be some other unsuspected producing-horizon at greater depth. The results from such drilling, whether success or failure, are known to everyone, and all of us use this information to guide our own future exploration. In such a system there is plenty of opportunity for surprises. Many unanticipated oil fields are found this way, but, of more importance socially, the whole country is thoroughly explored and much new prospective territory is revealed where more intelligently placed tests can then be undertaken.

Is it then a superior quality of geologic attainment that has enabled Americans to find oil? Emphatically, no. Russian geologists, British, German, and French geologists are generally better informed, more thoroughly trained in fundamental concepts and often more brilliant than American geologists. Americans excel only in applying an adequate knowledge of geology more effectively and more extensively to the search for oil. In a country where class distinctions bar intimate collaboration between scientist and laborer as fellow workers, or where government prohibitions or social habit deny the wildcatter freedom to drill exploratory wells throughout the land, the American method of oil-finding would fail.

Americans have drilled almost a million wells in the United States, more than 200 thousand of which have been dry holes. As long ago as 1921 this country drilled 34 thousand wells in a single year. No other country ever approaches such records. Eugene Stebinger calculates that whereas we have drilled one exploratory test for each 12 square miles of sedimentary rocks the rest of the world has drilled only one exploratory test for each 1100 square miles of sedimentary rocks. On the basis of exploratory wells drilled, the search abroad has been only about one per cent as thorough as ours. Against our record of one well to each 12 square miles of sedimentary rocks, the British Empire has drilled one well to each 1,000 square miles: Russia, one well to each 2,000 square miles: Japan, one well to each 135 square miles: Germany, one to each 120 square miles. Is it strange in the light of this comparison that we have found more oil than all the other nations together?

Stebinger's unpublished analyses reveal that, in the more thoroughly explored parts of the United States, including the states of New York, Pennsylvania, West Virginia, Ohio, Oklahoma, and California, the density of exploratory drilling is equivalent to one dry hole for each two square miles, or so, of sedimentary rock area; the figure varies from an average of one dry hole to each 1.5 square miles in New York and Pennsylvania, combined, to one to each 2.7 square miles in California. As a result of this relatively thorough exploration from 1.1 per cent to 1.7 per cent of the total sedimentary area of these states produced oil.

The proportion of oil-producing area to total area of sedimentary rocks is relatively constant in each of these states, despite widely divergent geologic character. The inference is that with complete exploration by drilling — exploration, say, to a test-well density of one test well to each square mile— from 1 per cent to 2 per cent of the total area of the average reasonably favorable sedimentary basin would yield oil commercially. If this inference is valid the United States as a whole, with 0.9 million square miles of favorable sedimentary-rock area should contain some 10 million acres of productive oil land. If exploration by drilling is finally carried out all over our country with the thoroughness our more developed states have enjoyed, and if we assume per acre yields equal to the average for the area already proved for production in the United States, this country should ultimately yield at least 100 billion barrels of oil, including the 44 billion barrels already discovered. On the same basis the rest of the world, with some 6 million square miles of known reasonably favorable sedimentary-rock area, would ultimately produce some 600 billion barrels of oil, including the 37 billion barrels already found. Such figures suggest the size of our ultimate reserve of oil in the earth.

It is not any apparent dearth of oil in the earth that need give concern to society. Society, if a geologist may look over into the field of the social scientist, might better look to its own failures than to the failures of nature. The probable ultimate oil resources of the earth, made available and freely

TABLE 1

EXPERIENCE IN EXPLORATION FOR OIL IN UNITED STATES

Year	(1) Thousand Dry Holes	(2) Major Fields Disc'd	(3) Billion Barrels Disc'd	(4) Dry Holes Per Major Field	(5) Thousand Bbls. Discovered Per Dry Hole	(6) Average Weighted Price for Oil (Dollars)
1920-'22	18.2	20	3.34	910	180	2.30
1921-'23	16.9	25	3.34	680	200	1.75
1922-'24	17.1	26	2.51	660	150	1.60
1923-'25	18.3	28	2.20	650	120	1.70
1924-'26	20.1	40	4.33	500	220	1.90
1925-'27	21.9	43	4.57	510	210	1.80
1926-'28	22.3	54	6.72	410	300	1.60
1927-'29	22.3	47	5.45	470	240	1.35
1928-'30	21.9	41	10.01	530	460	1.30
1929-'31	18.5	31	8.71	600	470	1.10
1930-'32	14.1	20	7.30	670	510	0.90
1931-'33	10.6	22	2.20	480	210	0.70
1932-'34	11.2	26	2.65	430	260	0.80
1933-'35	12.6	42	3.98	300	320	0.90
1934-'36	14.5	51	5.15	280	360	1.05
1935-'37	16.6	58	5.06	290	300	1.10
1936-'38	17.7	56	4.85	320	270	1.15
1937-'39	18.8	46	3.92	410	210	1.15
1938-'40	19.0	27	2.61	700	140	1.10

distributed, should meet humanity's needs for 300 years to come. The great difficulty is to establish the social conditions which will enable men to find and develop these ultimate resources. Our present proved reserves, world-wide, are hardly equal to 20 years' requirements. Americans, who have found the earth's oil in the past, face an even larger responsibility for the future. Other nations simply have

TABLE 2

EXPERIENCE IN EXPLORATION FOR OIL IN TEXAS

Year	(1) Thousand Dry Holes	(2) Major Fields Disc'd	(3) Billion Barrels Disc'd	(4) Dry Holes Per Major Field	(5) Thousand Bbls. Discovered Per Dry Hole
1920-'22	4.8	7	0.47	690	98
1921-'23	4.0	8	0.58	500	145
1922-'24	4.5	9	0.58	500	130
1923-'25	5.2	14	0.91	370	175
1924-'26	6.7	22	2.37	300	350
1925-'27	7.9	26	2.63	300	330
1926-'28	8.9	27	2.81	330	320
1927-'29	9.4	21	1.68	450	180
1928-'30	9.5	19	6.62	500	700
1929-'31	.8.1	13	6.98	620	860
1930-'32	6.3	8	6.44	790	1010
1931-'33	5.1	9	1.36	570	270
1932-'34	5.6	15	1.86	370	330
1933-'35	6.4	26	2.57	250	400
1934-'36	7.2	32	3.05	220	420
1935-'37	7.8	30	2.38	260	310
1936-'38	7.9	28	2.20	280	280
1937-'39	7.7	18	1.65	430	210
1938-'40	7.1	11	1.10	640	150

not found oil and in a number of these nations
American methods of exploration are barred by
existing social usages. Yet American methods, machinery, initiative, and, above all, American freedom of enterprise appear to be indispensable to
the task of finding oil in the earth.

The rate of discovery of oil fields in the United
States has declined somewhat in recent years.
Tables 1 and 2 show for the United States and for
Texas respectively the actual number of major oil
fields, and the estimated volume of oil discovered
over successive three-year periods during the last 21
years. For the purpose of this study, major oil fields
include only those fields whose ultimate production
will exceed 20-million barrels (a week's supply for
the nation at current rates of consumption). Such
fields account for nearly 80 per cent of the total oil
discovered in the United States during the period
under review. Estimates of the total volume of oil
discovered, as shown in the tables, include all fields,
minor as well as major. In these tables every major
field is credited to the year in which it was discovered, even though its major character may not
have been revealed until afterward. Every extension
to a known field or revision of the estimates of reserves in a known field is credited back to the year
of discovery, not to the year in which the extension
or revision was made.

The number of new oil fields found by wildcatting in a given period, depends on the effort devoted to exploration and on its effectiveness. A dependable measure of the effort is the number of dry holes drilled.

We may gauge the effectiveness of the wildcatting effort either by the number of oil fields, or by the number of barrels of oil, discovered. Perhaps the number of fields discovered is the better criterion because each significant discovery is an achievement of equal rank in oil-finding whether the field be large or small; whereas, measured by its oil content, a single field such as East Texas, with original reserves estimated at from four to five billion barrels, appears to be as important as forty or fifty ordinary hundred-million-barrel oil fields.

For both the United States and the state of Texas the effort devoted to oil-finding, as measured by the dry holes drilled, has been much less on the average over the past decade than it was in the 1920's. The effort has increased with the rise in prices since 1934 but is still less than it was from 1924 to 1930. Even more precipitate has been the decline in the success of the finding effort both in the United States at large and in Texas. Whether measured by the number of major fields discovered or the size of the reserves in the newly discovered fields the record shows a lower discovery rate from about 1934 forward. The period 1934-'36 yielded roughly double the number of new major fields and double the new reserves found in the period 1938-'40. Present rates of discovery, measured by either index, are lower than the average level for the last twenty years. In Texas, approximately three times as many wells were drilled to find a major oil field in the period 1938-'40 as in the period 1934-'36. For the United States as a whole the effectiveness of exploratory effort decreased nearly as much.

The item of dry-hole cost per barrel of oil found has mounted in recent years. This cost is only one item in the cost of finding oil and it is not the major item; but it has grown in relative importance until it has come to be a larger item than most of us realize. The cost of drilling a dry hole is probably greater today than it was ten or twenty years ago because wells are constantly increasing in depth. But even if we assume that the average cost of a dry hole has been constant the item of dry-hole cost per barrel of oil discovered in Texas was nearly seven times as much over the last three years (1938-'40) as it was during the heyday of oil-finding (1930-'32), and more than twice the average for the 21-year period, 1920-'40. For the United States at large dry-hole costs in the recent period (1938-'40) were more than three times as much as they were in the most successful period (1930-'32), and more than twice the average over the whole period, 1920-'40.

Mounting dry-hole costs make us reflect that the American method of finding oil fails unless we have always on the job a multitude of independent wild-catters. In no other way can we maintain the activity of drilling the thousands of wells we require each year to guide our co-operative search intelligently. The fundamental factor in stimulating exploration has always been an oil price high enough to return to the small producer the cost of his exploratory dry-holes as well as a reasonable profit on his current oil sales. Oil prices have been too low in recent years to stimulate wildcatting. An inadequate price for oil is certain in time to destroy many small operators and thus to sacrifice the most precious heritage our experience as oil finders has bequeathed to us—our multitude of itinerant wildcatters.

This inquiry into recent discovery rates in the United States and the detailed review of finding experience involved in it has diverted us from our larger question, "Who Finds Oil and How?" In summary it may be said that it is Americans who find oil in the earth and that their methods of finding involve a thoroughly coordinated geological research, the final step in which has been intensive exploration by drilling of actual test wells. Exploration by drilling is conclusive and over the life of the American industry it has not been prohibitively expensive, although at intervals, such as the last half decade, when the search has lagged for want of adequate stimulus, costs of finding mount. The important conclusion is that the method is effective: where we put it to the test we really find oil. Demonstrated also is the amazing abundance of oil in the earth if only we try diligently and intelligently to find it.

IV

WHOSE OIL IS IT?

IV

WE HAVE reviewed the nature of oil in the earth's crust, the character of its environment, the types of rocks in which it accumulates, and the manner in which men have contrived to learn just where it is concealed in remote places within the earth. We have now to inquire how it is that oil so resolves the affairs of men that they strive everywhere to reduce it to possession; and to observe how the several nations have fared in the apportionment among themselves of the earth's oil resources.

Oil has long been indispensable to man: for the caulking of boats, for the worship of the gods, as a medicine, as a lubricant and as a source of light and heat; always oil has found a ready market, in ancient times as in modern, alike in Babylon and in World War II. Today oil is at once the most vital necessity and the most highly prized luxury of society. It is the raw material for a thousand products: dyes, chemicals, drugs, explosives, fabrics, plastics, tires, lacquers, solvents and even edible fats are now derived from it. As fuel and lubricant it moves us over the earth at incredible speed in motor car, airplane, train and ship; it carries our burdens in truck, train and boat; it smelts our metals from their ores; it ploughs our fields; harvests our crops; heats our homes; lights our lamps; cooks our meals, and generates power for mill and factory and mine. Oil is the staff of life!

But oil only became of prime importance when automobiles replaced horses on land and ships at sea substituted oil for coal as fuel. These two changes were speeded by war; a statement made during World War I by Winston Churchill fixes the date. Mr. Churchill, as First Lord of the Admiralty, spoke of the difficulties of converting the British Navy from coal to oil fuel:

"Better ships, better crews, higher economies, more intense form of war power—in a word, mastery itself was the prize of the venture. A year gained over a rival might make the difference. Forward, then!

"During the whole of 1913 I was subjected to an ever-growing difficulty about the oil supply. We were now fully committed to oil as the sole motive power for a large proportion of the fleet, including all the newest and most vital units.

"There was great anxiety on the Board of the Admiralty and in the War Staff about our oil-fuel reserves. The Second Sea Lord, Sir John Jellicoe, vehemently pressed for very large increases in the scales contemplated. The Chief of the War Staff was concerned not only about the amount of the reserves but about the alleged danger of using so explosive a fuel in ships of war. Lastly, Lord Fisher's Royal Commission, actuated by Admiralty disquietude, showed themselves inclined to press for a reserve equal to four years' expected war consumption.

"The war consumption itself had been estimated on the most liberal scale by the Naval Staff. The expense of creating the oil reserve was, however, enormous. Not only had the oil to be bought in a monopoly-ridden market, but large installations of oil tanks had to be erected and land purchased for the purpose. Although this oil-fuel reserve, when created, was clearly whether for peace or war, as much an asset of the State as the gold reserve in the Bank of England, we were not allowed to treat it as capital expenditure: all must be found out of current estimates.

"At the same time, the Treasury and my colleagues in the Cabinet were becoming increasingly indignant at the naval expense, which it might be contended was largely due to my precipitancy in embarking on oil-burning battle ships and also in wantonly increasing the size of the guns and the speed and the armour of these vessels. On the one hand, therefore, I was subjected to this ever-growing naval pressure, and on the other to a solid wall of resistance to expense. In the midst of all lay the existence of our naval power."

Thus the First Lord of the Admiralty reflected the pressure under which Britain at that time felt driven to assure herself of adequate supplies of oil. Under this compulsion she entered into negotiations which soon culminated in British possession of the oil fields of Persia which, as we have previously observed, constitute the bulk of the proved reserves at the Near East end of the earth's oil axis. For the purpose of holding this reserve, Britain created the Anglo-Persian Oil Company in which the government itself held a majority of the stock. This company, in collaboration with the Royal Dutch Shell, an older corporation organized and still controlled by British and Dutch citizens for the purpose of developing the extensive oil resources of the Dutch and British East Indies, later came into possession of other important reserves when they secured from the King of Irak, a new state under British mandate in the territory formerly known as Mesopotamia, in the Near East, the famous oil lands of Mosul. In this enterprise American and French capital came eventually to share as minority partners.

Great Britain was still uneasy about its oil resources even after its acquisition of the rich oil fields of Persia, or Iran, as the present Shah has renamed his country. In May, 1920, Sir Auckland Geddes, the newly arrived British Ambassador to the United States, speaking before The Pilgrim Society in New York, complained that the British Empire had only 4.5 per cent of the world's oil production, including Persia, whereas the United States, he declared, had more than 80 per cent.

From time to time our own government has manifested an interest in oil reserves outside our national boundaries. In 1920, when the Allies were engaged in dividing among themselves the oil resources of the territory taken from Turkey during the World War and later segregated as the British Mandate of Irak, our Department of State, in keeping with its long-established policy of the Open Door, insisted vigorously that in the disposal of these oil resources Americans be given consideration among other nationals. Because of this and other pressures the British finally accorded American capital a 24 per cent participation in the concession covering the oil resources of Irak.

Again, in 1926, our government, through the Federal Oil Conservation Board, urged upon the American oil industry a policy of acquisition of foreign oil properties. A report of the Board published at that time contains the following statement:

"The fields of Mexico and South America are of large yield and much promising geologic oil structure is as yet undrilled. That our companies should vigorously acquire and explore such fields is of first importance, not only as a source of future supply, but supply under control of our citizens. Our experience with the exploitation of our consumers by foreign-controlled sources of rubber, nitrate, potash and other raw materials should be sufficient warning as to what we may expect if we shall become dependent upon foreign nations for our oil supplies. Moreover, an increased number of sources tends to stabilize price and minimize the effect of fluctuating production."

Responding to this suggestion from their government American companies did acquire and explore concessions in Mexico and South America. Moreover, they developed valuable oil resources in these countries. Unfortunately, after oil had been discovered on them and the previously trackless jungles had thus been made valuable, some of these properties were expropriated by the very governments which had granted the concessions. The issue was: whose oil is it?

Disturbed world conditions have quenched most of the interest in oil resources outside our own national boundaries which formerly animated our government. Surely this interest should now be rekindled; not in the selfish interest of our own country alone but in behalf of the entire Western Hemisphere, that "commonwealth without empire" which we aspire to erect in the New World. There to the south of us lie the great undeveloped oil resources of our sister nations; Mexico, Colombia, Venezuela, Ecuador, Peru and Brazil. Here in American free enterprise are the skills, the capital, the organization and the techniques best suited to explore and develop these resources. But whose oil is it? The social and economic advancement which will bring to our Latin American neighbors the highest possible standards of living requires the development of these oil resources. The need is imperative. Shall we not, under agreements that abundantly safeguard the national interests of each of these countries, place our special talents at their service in making available to all of the nations of the earth the oil resources of the earth?

Whatever the facts may have been when the British Ambassador in 1920 credited the United States with 80 per cent against Great Britain's 4.5 per cent of the earth's oil, the situation today is more nearly balanced. We have already observed that most of the 81 billion barrels of oil so far discovered on earth is concentrated in the Near East, where Europe, Asia, and Africa meet, and in the Gulf of Mexico-Caribbean Sea region between North and South America. Nearly two-thirds of the total proved oil reserves of the earth and fully 80 per cent of the reserves outside of the United States are concentrated in these two regions. At the present time British companies, in which the British Government itself participates, hold about 80 per cent of the proved reserves in the Near East, excluding Russia. A British-Dutch company holds a somewhat smaller proportion of the proved reserve in the British and Dutch East Indies. Americans hold most of the balance of the oil both in the Near East and in the Orient.

Of the proved reserves in the Gulf of Mexico-Caribbean Sea region, excluding the part that lies in the United States, Americans hold about three-quarters, while British companies hold most of the balance. In addition, the British have built up respectable reserves in India, and occupy a modest position elsewhere in the world, including a considerable volume of production within the United States itself. Americans, of course, hold practically all of the proved reserves in the United States.

Summing up the situation, it appears that of the earth's proved oil reserves outside the United States, the British now hold about 50 per cent, Americans 25 to 30 per cent and Russia around 20 per cent.

If proved reserves within the United States itself be included then Americans have about 60 per cent, the British 25 per cent and Russia 10 per cent of the proved reserves of the world.

It is clear from these figures that, outside of Russia, the proved oil reserves of the world are almost entirely held by Americans and British. It should be noted that to "hold" in the sense here intended does not mean ownership. American and British corporations, having negotiated concessions or contracts with the owners of this oil, who are usually national governments, have discovered and developed the oil and now sell it on the world's market under the terms of their contracts. It should be noted also that the estimates refer only to proved reserves. The undeveloped potential resources of oil, which are likely to be greater than the proved reserves, may eventually come to reside in other proprietors.

Russia is at once a proprietor and an owner of oil in the earth. The profusion of her resources was indicated in a previous lecture. Still largely undeveloped even after three-quarters of a century of exploration Russia's potential resources in oil surpass those of any other nation. More completely than any other nation Russia is self-contained as to her future requirements for oil. Imperial Russia had exported important quantities of oil. Since the revolution exports have ceased. Future policy will probably earmark all of her production for her own needs.

Germany long ago realized the value of oil resources as national assets. In June, 1914, the Deutsche Bank of Berlin finally consummated with the Turkish Government a trade which had been under negotiation for more than 15 years. By the terms of this trade, as a part of the Berlin-to-Bagdad Railway concession, German, British and French associates obtained rights to develop the oil resources of the "vilayets of Mosul and Bagdad" in Mesopotamia. These vilayets are included in the modern kingdom of Irak, a rich oil-producing section of the Near East, but the German interest (said to have amounted to twenty-five per cent) had been eliminated when peace was finally concluded after the close of the World War.

Germany had never developed worth-while oil production within her own boundaries, although many small oil fields had been discovered there. After the World War when the nation set about to re-establish itself it began at once the attempt to synthetize oil from lignite coal. This attempt succeeded technically but costs were high and the output remained too small to meet national demand. Meantime her internal production of natural oil had increased to considerable proportions but, even so, total internal production remained inadequate and Germany continued to be an importer of oil up to the outbreak of World War II.

France, Italy, Spain and the smaller nations of Europe, except Roumania, are all importers of oil although France, Italy, Hungary, Austria and Poland all produce some oil. France owns a 25 per cent interest in the concession covering the Kingdom of Irak but her share of the output does not meet her full needs. All of Africa, except Egypt which is a considerable producer, depends entirely on imports. In the Orient, Japan and India each produce oil but have to import most of their requirements. China depends wholly on imports of oil although promising undeveloped resources are known to exist inside her boundaries, particularly in the northwest provinces. Australia and New Zealand produce no oil and their supplies, like those of the Orient in general, come largely as imports from sources in Sumatra, Java and Borneo, developed by British, Dutch and American producers.

In South America, Brazil, Chile, Paraguay and the Guianas import all the oil they use, while Argentina, Bolivia and Ecuador each produces only about 50 per cent of her requirements. All of Central America and the West Indies, except for a small indigenous production in Cuba, import all their oil. Canada, with potentially great undeveloped oil resources, imports most of her requirements.

WHOSE OIL IS IT?

The oil-exporting countries of the world other than the United States are Venezuela, Colombia, Mexico, Peru and Trinidad in the Western Hemisphere; Roumania in Europe; Iran, Irak, Arabia and Egypt in the Near East; and Borneo, Sumatra, Java and Burma in the Orient.

In the natural distribution of the great accumulations of oil in the earth there is a striking accommodation to the principal centers of consumption on the earth. In the Western Hemisphere, North and South America both draw conveniently upon the great potential resources of oil in the Gulf of Mexico-Caribbean region. The tremendous reserve in the Near East should serve Europe and Africa adequately and efficiently. Russia, more abundantly than any other nation, has potential supplies within her own borders in proportions commensurate with her future internal needs. In the Orient, China, Japan and Australia might take their supplies naturally from the Dutch and British East Indies. India and Burma have great oil resources of their own little explored as yet and in addition to these they can readily draw upon both the Near East and the Orient. No really populous region of the earth is remote from abundant oil resources. Nature appears to have provided every nation with a convenient source of oil. Yet oil is not in fact ratably distributed among the nations today. The obstacles to fair distribution of oil are man-made, however;

they are not chargeable to Nature. Wars, embargos, policies of national self-containment, cartels, exchange restrictions, tariffs, and other trade barriers; these are some of the impediments to the free flow of oil to the several nations.

The fourth principle of the Atlantic Charter recognizes the needs of all nations for strategic materials and minerals:

> to "further the enjoyment by all states, great or small, victor or vanquished, of access on equal terms, to the trade and to the raw materials of the world which are needed for their common prosperity."

Any commonwealth of the nations of the earth needs as a cornerstone of its foundation a wise policy for the full development of oil in the earth.

It is significant that in the United States where the development of oil resources is further advanced than anywhere else on earth, the peoples of all nations have been permitted to engage freely in the production of oil. A certain Texas oil field where adjacent properties were administered by an Armenian, a Japanese, and an American, respectively, exemplifies the free economy that has accomplished the most effective search for oil fields on record. If it were possible to extend American oil-finding practices to the rest of the world and to establish in other countries the same freedom for vigorous exploration that has characterized our own way of life, the world's oil resources would soon be developed as completely as our own have been.

A wise oil policy for the United States and for the world at large would include a veritable crusade to develop the oil resources of the earth. Most of the world requires far more oil than it has ever had. No industrial nation can exist without oil. Oil turns the wheels of all modern industry. The world's merchant marine floats on oil. Nowhere on earth are standards of living sustained at high levels except where they rest upon cheap transportation— "the greatest economic and social revolution which has ever taken place"; and cheap transportation today depends completely on oil. Abundance of oil for every nation is an indispensable preliminary to "freedom from want" for every nation. Without abundant oil there can be no high standard of living; with abundant oil high standards of living can be erected in other countries just as they have been in the United States.

A wise oil policy for all nations would include the following principal objectives:

(1) The rapid development of the oil resources of the earth to the point where all nations are assured access to adequate supplies "for their common prosperity"; and to the point where proved reserves are of sufficient volume to meet future world needs over a reasonable period.

(2) The opportunity for participation by the citizens of all nations in the development of the oil resources of the earth under the conditions of free enterprise, preserving to all nations, however, perfect freedom to formulate their own legislative control for the development of their own resources, within the scope of their established philosophy of government and jurisprudence.

(3) The orderly production of oil from the oil fields of the earth without waste, controlling withdrawals of oil uniformly to the optimum rate for each oil field, conserving both oil and reservoir energy with utmost efficiency.

(4) The efficient distribution ratably from each nation of the exportable surplus of oil, as determined by that nation, among other nations as needed.

These objectives can best be attained by spreading over the world the same free enterprise that has found and developed so successfully the oil resources of the United States.

The principal barriers to the rapid, complete development of the oil resources of the earth are:

(1) Government monopolies which place oil resources in the hands of a government agency and deny to all other citizens any right to explore or develop them.

(2) Laws which withhold from foreign citizens any right to engage in exploration or development of the oil resources of a country.

Oil monopolies have been established by Russia, Brazil, Mexico, Japan and Bolivia. Spain, Argentina and Chile maintain semi-monopolies which hamper but do not entirely prevent operations by private individuals. The laws of a number of other nations, and of colonies and mandates discriminate against foreigners in prescribing conditions under which oil resources may be explored and developed.

Obviously no other nation should be expected to adopt our specific laws or regulations for its oil industry. Each nation will maintain intact its own laws and customs. The government ownership of all oil resources, for example, which is common among nations, would nowhere be molested. Each nation would control completely its internal economy, including the export and import of oil. It is to be hoped, however, that government monopolies would be relinquished sufficiently to permit private citizens to engage in the search for oil, and that the laws would be modified sufficiently to permit foreign citizens to participate in the oil industry on the same terms accorded to nationals. If free enterprise were welcomed everywhere in the search for oil in the earth, society at large would be greatly served.

93

Americans are accused of wasteful practice in producing oil. In our haste to discover new oil fields we have drilled exploratory wells that need not have been drilled if we had been content to wait until the revelations of wells already drilled had been more fully interpreted. As a consequence our "finding costs" are higher than they would have been had we proceeded more cautiously. Nevertheless, our overall finding costs are not exorbitant; and with a less aggressive campaign of exploration, which might have spared us some useless drilling, we would have failed to discover all the oil fields we have found.

Again, in our haste to recover the oil from our oil fields we have permitted the associated natural gas to escape into the air and go to waste. We know today that natural gas is potentially gasoline. We can transform the one to the other. We should have conserved more of our natural gas in the past. In the modern oil field the natural gas that comes to the surface of the ground with the oil is "recycled"; that is, it is recompressed and forced back down the wells to be stored in the reservoir whence it came. Recycling plants increase the initial investment required for oil-field operation but in many instances they permit a lower overall cost of producing oil and in addition they conserve the gas.

Our rapid withdrawals of oil and waste of gas from oil fields in the United States have also depleted the pressures under which oil and gas are originally confined in the underground reservoir; the so-called "reservoir energy" which should be employed efficiently to bring the oil to the surface of the ground. In this way we have added to our "lifting costs" and at the same time subtracted from the total volume of oil recoverable from a given oil field, since with the pressure in the reservoir prematurely dissipated much oil is left behind in the sand, unrecoverable except at a higher cost.

Our daily production of oil in the United States amounts to nearly 4 million barrels from a total estimated reserve in all fields of about 19 billion barrels. This gives our average oil field a future producing life of less than 13 years. Yet our experience indicates pretty conclusively that for maximum efficiency the average oil field should be operated so that its producing life will not be less than about 20 years. Not more than about 5 per cent of the total contents of the average oil field should be withdrawn each year. The oil should be permitted to flow out of the reservoir only as fast as the associated water presses in to take its place. Control of withdrawals on this basis will maintain reservoir pressures practically undiminished throughout the producing life of the average field (assuming the associated gas is returned to the reservoir), and under this pressure practically all the oil will flow from the wells spontaneously, without the expense of pumping. The flow of oil may safely be restricted below this "optimum rate" without waste or harm but if the optimum is exceeded reservoir pressures decline and losses of oil result.

If we were to adhere to this optimum rate of oil production in the United States we should not produce more than about 3 million barrels a day from our present proved reserve; or, conversely, if we wish to produce efficiently as much as 4 million barrels a day from our domestic oil fields we should increase our proved reserve by new discoveries to about 30 billion barrels.

Excessive rates of production in American oil fields have been due in part to our law of property rights in oil and gas. The so-called "law of capture" has driven the owner of one property in an oil field to produce his oil as rapidly as possible in order to prevent it from escaping into the possession of his competitor on an adjacent property in the same field. Our laws regard oil and gas as "fugacious substances" which move about as wild game does from one property to another and belong to no one until they are "reduced to possession." The oil beneath one man's property may flow through the porous rocks of the reservoir to a well on an adjacent property; if it does so the adjacent owner has only to bring it to the surface and reduce it to possession to make it his oil. Thus each operator must bring all his oil to the top of the ground before his neighbor takes it away from him.

So under the spur of possible loss to his competitor the American oil producer has been driven to withdraw his oil from the ground more rapidly than the principles of conservation would dictate. This evil practice can be eliminated only by the unitization of oil fields; that is, the pooling and joint operation of all the separate properties in each oil field; the surrender of each divided interest in a common pool of oil and gas for an equivalent undivided interest. Once the oil and gas in an entire oil field are owned in common it makes no difference what well or what property on the surface produces a given barrel of oil; every joint owner shares in the proceeds from its sale.

The unitization of oil fields is difficult of accomplishment because it requires agreement and compromise as to the respective shares of different owners. Men do not agree readily on such matters and unitized oil fields are therefore not common in the United States. Instead there has been established under state laws an expedient known as proration which, while it does not eliminate the evils of the law of capture, mitigates the worst effects. Under proration each producing property in an oil field is operated in unison with every other property in that field, conforming to individual quotas of allowable production fixed for it by the conservation authority of the state. These quotas are so regulated as to do equity between competing producers and at the same time to restrict total withdrawals from the oil field to a rate low enough to maintain reservoir pressures and so avoid physical waste. In other words the quotas fixed by proration authorities establish the optimum rates of withdrawal for each property in the field. Proration under state laws has enabled American oil producers to operate their properties more efficiently and to observe more strictly the principles of conservation. The principal oil-producing states have correlated their individual proration programs in the interest of conservation through the Interstate Oil Compact, established with the approval of Congress.

The extravagance with which we have used our liquid fuels in the United States may also be viewed as a form of waste. We have never found it worthwhile to reduce the weight of our automobiles to the minimum nor to increase the efficiency of our internal combustion engines to the maximum. Gasoline has been so cheap that efficiency hardly justified itself. We have been willing to burn fuel oil instead of coal even in stationary boiler plants; the fuel oil has been cheaper. We have been content to recover 15 or 20 gallons of gasoline from a barrel of oil when we might have recovered (at a slightly higher unit cost) 25 or 30 gallons from the same barrel.

One consoling reflection emerges from a consideration of these data, however; if we should find it necessary to get along on, say 3 million barrels of oil daily in the future instead of the nearly 4 million barrels we are using currently, we can easily do so. We can cease to burn oil where coal might better be burned; we can convert more of the average barrel of oil into gasoline; and we can design motors and motor cars to make more efficient use of gasoline and oil. If we do all these things, 3 million barrels of oil daily will meet our national requirements for many years to come.

Students of the American oil industry often deplore the hasty and uncontrolled depletion to which our early oil fields were subjected. They are inclined to think "there ought to be a law" to prevent these excesses. Why should not unitization of oil fields be made compulsory, for example, even if the competing operators are unwilling to pool their individual properties? The answer is that compulsory unitization violates fundamental property rights; moreover, it must impose restrictions on the oil-producing industry which would stifle the very freedoms upon which its entire achievement rests. We can insist upon unitization and other checks to individual action and by so doing we can reduce both waste and costs. But if we establish these controls, we shall fatally handicap American oil finding. The wastes and the excess costs we would eliminate are not large. They are insignificant in comparison with the benefits that have come to us as a result of our bountiful proved reserves of oil.

To the credit of the American oil industry it must be acknowledged that it has made available to the consuming public adequate supplies of liquid fuel at low cost. We have noted that with only 15 per cent of the favorable hunting-ground within her boundaries, the United States has found more oil than all the rest of the world. No other nation has developed its oil resources with such great advantage to its industrial, social and national life. American transportation facilities both for passengers and for freight, over land, on sea, or through the air, have always enjoyed generous supplies of gasoline. And the cost of this gasoline, despite a tax roughly equal on the average to the total wholesale price at the refinery, is far below the world price for gasoline. America's industrial supremacy and high standard of living rest upon this foundation: an abundance of energy at low cost, available to everyone, everywhere, in the convenient, concentrated form of liquid fuel. In our achievement of these signal advantages, we are a generation ahead of other peoples. Cheap, convenient, mechanical power, subject to the call of every citizen, everywhere, is the one feature that has distinguished our social economy over the last two decades from that of every other nation.

WHOSE OIL IS IT?

One of the factors widely alleged to have driven the "have not" nations, Germany, Italy and Japan to ventures in military conquest is their urgent need for oil. There is no validity in any claim by these nations that they had to go to war to get the oil required for the maintenance of their social and industrial orders. Oil in the earth is too lavishly and too indiscriminately distributed. A fraction of the effort that pours into war, were it directed effectively to the exploration of the crust of the earth, could hardly fail to reveal important oil resources in each of these countries. In none of them has anything like adequate exploration for oil been attempted. Germany spreads over a great sedimentary basin, studded with oil-impregnated salt domes. The rocks in Italy are characterized by a conspicuously petroliferous series of shales and by geologic structure favorable to oil accumulations. Japan includes large areas underlaid by thick sections of the same rock layers that produce oil to the southward in Sumatra and to the northward in Sakhalin. These rocks are marked by copious seepages of oil.

In the light of American experience in oil-finding throughout the earth, it is difficult to escape the conviction that other nations might have enjoyed similar advantages from abundant, cheap, liquid fuels had they sought as diligently for oil in the earth within their own boundaries as we have in the United States. Surely Germany, for example, could have developed large oil reserves in her own territory had she pursued American methods of exploration with as much vigor as she has waged war. The nations surrounding her, including Hungary, Jugo-Slavia, Poland, even Italy and France, are attractive territory to the American wildcatter. Roumanian production in a free economy would almost certainly expand materially. These nations could discover great oil resources within their own domains if they maintained free economies and adopted American methods. Japan has left unexplored so much promising territory in the Far East, including her homeland, that she can scarcely contend that she is driven to the conquest of the Netherlands East Indies by her imperative need for the oil British, Dutch and Americans have developed there.

Much more oil remains to be found in the United States in addition to the present proved reserve. Under the impetus of higher prices and the momentum of new discovery itself, once it is resumed (discovery in the United States has faltered in recent years), this additional oil will be found. Nevertheless, however much oil remains to be discovered, the supply available at low cost must finally be exhausted, unless our need for liquid fuel and lubricants ceases meantime. We have taken far more oil from our part of the earth than other great nations have taken from theirs. The prospect of exhaustion of our oil reserves has alarmed many people. What are we to do when our own sources of oil fail?

One of the first steps to be taken whenever scarcity threatens in oil supply is to develop fully the sources most adequate and best adapted to serve the principal centers of consumption; another is to distribute equitably and efficiently the products from these sources. These principal sources we have already identified as the Near East end of the earth's oil axis for Europe and Africa; the Gulf of Mexico-Caribbean Sea end of this axis for the Western Hemisphere; and the Far Eastern portion of the Pacific oil province for the Orient. Whenever these steps are taken scarcity will be put off many decades into the future. Judging from the experience of Americans in their exploration for oil over the earth, the undiscovered oil resources must aggregate hundreds of billions of barrels. The ultimate supply should be adequate for generations to come.

But these great potential sources of oil, as yet hardly scratched by the drill, must be explored and developed and their output must be equitably and efficiently distributed. No individual nationalistic ambitions can be permitted to interfere. A truly Good Neighbor Policy between this country and Latin America, for example, would come in time to envisage great imports of oil into this country from tremendous new oil reserves still undiscovered in Brazil, Peru, Ecuador, Colombia, Venezuela, Central America, and Mexico. Latin America would take reciprocal exports from this country and would welcome American engineering talent and financial resources to assist in the formidable task of converting vast tropical wastelands into healthful, productive oil fields for the common benefit of all the nations of the New World. Similar freely-functioning exchanges must be established for the full development of the oil resources of the Near East and of the Orient, and for the unrestricted distribution of products therefrom.

In the distant future, however, even these great sources of oil in the earth, together with all the supplemental supplies from subordinate sources over the earth, must begin to fail. Whenever this happens, our social economy, gradually and without great difficulty, will substitute synthetic oil for earth oil. The techniques are already available. Only a somewhat greater cost prevents synthetic oils from taking their place in the market today. Indeed, as is well known, Germany has relied largely on synthetic oil to meet her needs recently even under the emergency of war. Other nations can do the same whenever the necessity arises.

These reflections throw doubt on the widely accepted idea that wars can be prevented in the future by denying to possible belligerents access to resources such as oil. It is already possible for the great powers, driven by national peril, to build up stocks of synthetic fuels and lubricants adequate for extensive military operations. The disease of war requires a broader, more comprehensive remedy, not so readily nor so simply devised.

The whole family of hydrocarbons we know as oil in the earth can now be synthesized, as we have already observed, from elemental carbon and hydrogen from whatever source. Our own nation has an abundance of raw materials wholly suitable for this purpose. We would use natural gas, or tar sands, or oil shales, or coal which we possess in aggregate volume equivalent to thousands of billions of barrels of oil. Whenever the occasion arises we too can do readily enough what Germany has already done: make our oil synthetically. The significant difference from our present practice will be the greater cost. Yet the added cost will not be intolerable; to manufacture gasoline synthetically will cost little more than the present sales price of gasoline plus the average tax on gasoline. In other words, if gasoline made from natural gas, or oil shales, or coal paid no tax it should be available to consumers at the point of manufacture at a cost but little higher than the present price of gasoline, including tax.

Before our oil in the earth is exhausted, then, we shall be making liquid fuels and lubricants synthetically from other raw materials which Mother Nature has stored in the earth for our use. In the first stages of this substitution of synthetic for natural oils we shall use as our raw materials natural gas, tar sands, oil shales, or coal. But eventually, unless meantime we have solved the problem of appropriating the internal energy of the atom at tolerable costs, we shall come to draw currently upon the ultimate source of our energy, sunlight; if not directly, then through the intermediate step of growing plants.

Thus in the end shall we free ourselves of our present dependence on oil in the earth!

INDEX

Accumulation of oil in anticlines, 23, 24, 25
  in closed systems, 12
  geologic structure related to, 23
Age of oil-bearing rocks
  Amazon River, head waters, 35
  Arabia, 34
  Gulf of Mexico-Caribbean Sea region, 34
  Iran, 33
  Iraq, 33
  Orinoco Basin, Eastern Venezuela, 34
  Russia, 34
  United States, 30, 31
Alpha radiation, origin of oil, 17
Amari dolomite, Iran and Iraq, 34
Amazon River region, oil production possibilities, 35
Americans, as oil finders, 56
  exploration work of, 56
Anticlinal structure, 23, 24, 25
Arctic, oil possibilities, 41-44
Athabaska Tar sands, 41
Australia, absence of oil, 46

Bacterial action, origin of oil, 17
Black Sea muds, 16
British Empire, oil production, 82
  wells drilled, 67

Caribbean region, oil production, 32, 34
Catalytic cracking, 17
Central America, potential commercial oil fields, 36
Churchill, Winston, 30
Coal, United States reserves, 48
Colombia, prolific oil fields, 36
Concessions, American, 83
Cracking, 12, 17

Deformation, oil and gas dispelled by, 23
Discovery of oil fields, 69, 70
Distribution, geographic, of oil, 89
Domes, as oil traps, 20
Dry holes, 72
  cost of, 73
  drilled in U.S., 1920-1940, 69, 71
  drilled in Texas, 1920-1940, 70, 71

Electric well log, 64, 65
Eldorado field, 7
Enterprise, American, 55-59
Exploration, American, 56
  American methods, 66
  factors causing American success, 57

INDEX

principles of, 25
Russian methods, 66
United States, 71

Federal Oil Conservation Board, 83

Gas, economic importance of, 94
in oil, 94
products of, 94
Germany, oil production possibilities, 101
oil reserves, 101, 102
wells drilled, 67
Gulf Coast oil fields, 21

Haworth, Erasmus, 5
Hydrocarbon family, 10
Hydrocarbons, source material, 18, 19
Hydrogenation, origin of oil, 13, 18

Importers of oil, 88
Iran, oil fields, 33, 82
Iraq, oil resources, 33, 82
Italy, potential oil production, 101

Japan, oil production, 37, 67, 101
exploration for oil in, 102

Kansas, discovery of oil in, 49
oil production, 5, 9

Lind, S. C., 16, 17, 18
Logs, well, 64, 65

Malayan Archipelago, oil production, 37
Maracaibo Lake region, prolific oil fields, 35
Mediterranean region, oil production, 32
Mexico, prolific oil fields, 35
Migration of oil, 19, 20
Military conquests, relation of oil to, 101
Muds, euxinic, 16

Nemaha Mountains, 6
New Zealand, 46

Oil, discoveries of, 31, 39
economic importance of, 80
exploration for, 49, 52
exporters of, 89
habitat of, 21
price of, United States, 1920-1939, 69
search for, by Americans, 56
source materials of, 15
supply, ultimate, adequacy of, 103
synthesis of, 19
uses of, 79
Oil axis in earth, 31, 33
Oil-bearing rocks, age of, 33, 34
Oil Conservation Board, Federal, 83
Oil concessions, American companies, 83
German companies, 87
Oil development, barriers to, 92

INDEX

Oil discoveries, United States and world, 55

Oil fields, British possessions, 82, 83
  discovery and development, American enterprise, 59
  Eastern Venezuela, 24
  Gulf Coast, 21
  Iran, 33
  Iraq, 33
  major fields, 32
  major fields discovered from 1920-1940, 69, 70
  new world end of oil axis, 32
  rate of discovery tables, 69, 70
  transitory nature, 14

Oil industry, in totalitarian countries, 63
  in America, 63
  service to science, 65

Oil monopolies, 93

Oil reserves, estimates, United States and world, 33
  exhaustion of, 102
  United States, total reserves, 31
  world, control of, 85, 86
  world, excluding United States, 31

Oil shales, 41

Oklahoma, oil production, 5

Open Door policy, 82

Origin of oil, 15

Orinoco Basin, 24, 34

Pacific region, 36

Paleozoic rocks, oil from, 38
  in Pennsylvania, 30

Rocky Mountain front, 39
  total production from, 31

Persian oil fields, 81

Petroleum (see "Oil")

"Petroleum poles," 32

Policy suggested for nations, 91

Polymerization, 17, 18

Pools, character of, 11, 12

Production of oil, Kansas, 5
  Oklahoma, 5
  Paleozoic rocks, total, 31
  Pennsylvania, 30
  United States, 31
  world, excluding United States, 31

Rate of Oil Production in U.S., 95

Reserves of oil, proved, United States, 70, 85
  ultimate, 68, 69

Reservoir rocks, 22

Russia, potential production of, 44
  wells drilled in, 67

Sedimentary rock area
  favorable for oil, 47
  per cent oil bearing, 47, 48

Source beds, Arabia, 34

Source material of oil, 15, 18

Stebinger, Eugene, 47, 67

Structure, accumulation of oil, 22

Synthetic oil, manufacture, Germany and U.S., 106
  price of, 106
  use of, 105

INDEX

Texas, East, original reserves, 72
  Gulf Coast, early drilling, 24
  rate of discovery of oil fields, 71

Thermal reactions, origin of oil, 16

Traps, arches and domes, 20
  stratigraphic, 24

United States, area favorable for
    oil development, 30, 55
  area productive of oil, 68
  exploratory drilling in, 67
  oil per cent discovered in, 46
  oil fields, prolific, 34, 36
  oil fields, rate of discovery, 71
  oil reserves, 32

sedimentary rock area yielding
    oil, 46
synthetic oil, raw material, 106
wells drilled, 65

Venezuela, Eastern oil develop-
    ment, 24, 34

Wasteful practices, 94, 95

Weeks, L. G., 47

Well logs, 64, 65
  electric, 64, 65

Wildcatting effort, 72

Williston Basin, 39, 40